Safer, Stronger, Smarter:
A Guide to Improving School Natural Hazard Safety

Prepared by

APPLIED TECHNOLOGY COUNCIL
201 Redwood Shores Parkway, Suite 240
Redwood City, California 94065
www.ATCouncil.org

Prepared for

FEDERAL EMERGENCY MANAGEMENT AGENCY
Michael Mahoney, Project Officer
Andrew Herseth, Task Monitor
Washington, D.C.

APPLIED TECHNOLOGY COUNCIL
Jon A. Heintz (Program Executive, Program Manager)
Veronica Cedillos (Project Manager)
Ayse Hortacsu (Project Manager)

PROJECT MANAGEMENT COMMITTEE
Barry H. Welliver (Project Technical Director)
Suzanne Frew
William T. Holmes
Christopher P. Jones
Lori Peek
John Schelling
Thomas L. Smith
Edward Wolf

REPORT DEVELOPMENT CONSULTANT
Laura Dwelley-Samant

PROJECT REVIEW PANEL
Ines Pearce (Chair)
Jill Barnes
Victor Hellman
Andrew Kennedy*
Rebekah Paci-Green
Cindy Swearingen
* ATC Board Contact

WORKING GROUP
Lucy Carter
Shawna Bendeck
Scott Kaiser
Jacob Moore
Meghan Mordy
Katherine Murphy
Jennifer Tobin

June 2017

Notice

Any opinions, findings, conclusions, or recommendations expressed in this publication do not necessarily reflect the views of the Applied Technology Council (ATC), the Department of Homeland Security (DHS), or the Federal Emergency Management Agency (FEMA). Additionally, neither ATC, DHS, FEMA, nor any of their employees, makes any warranty, expressed or implied, nor assumes any legal liability or responsibility for the accuracy, completeness, or usefulness of any information, product, or process included in this publication. Users of information from this publication assume all liability arising from such use.

Cover images: FEMA (top left); David Welker (bottom left); Ellen M. Banner, *The Seattle Times* (right)

For sale by the Superintendent of Documents, U.S. Government Publishing Office
Internet: bookstore.gpo.gov   Phone: toll free (866) 512-1800;   DC area (202) 512-1800
Fax: (202) 512-2104 Mail: Stop IDCC, Washington, DC 20402-0001

ISBN 978-0-16-094130-6

The Cottonwood School
of Civics and Science
640 S Bancroft St
Portland, OR 97239

Foreword

Our nation's elementary and secondary school buildings contain the future of our country. Over 50 million students attend approximately 99,000 public elementary and secondary schools with an additional 5.2 million students attending close to 34,000 private schools (NCES, 2016). Parents send their children off to school every day with the belief and expectation that their children will be safe from natural hazards. Children not only have the right to an education; they also have the right to an education in a safe environment. However, in many parts of our country, school buildings are vulnerable to severe damage or collapse in the next earthquake, tornado, hurricane, flood, tsunami, windstorm, or other natural hazard and are therefore putting our children at risk. In particular, many of our nation's school buildings are older unreinforced masonry (URM) structures that are vulnerable to severe damage and collapse in the next earthquake, or are of lighter frame construction that is vulnerable to other types of natural hazards such as a tornado, hurricane, high winds, or flash flooding. Some schools are located in tsunami hazard zones without access to safe ground that can be reached within the expected tsunami warning time.

Schools are far more than a place for teaching children; they often serve as community centers. They are the places where the public votes for their future leaders and they often serve as a focal point for a community's social and cultural life, be it the Friday night football game or the location for evening community meetings. The loss of a school building can severely disrupt the fabric of a community.

School buildings also serve other critical functions within the communities where they are located. For example, they often serve as designated shelters for displaced families after a natural or manmade disaster. Even when they may not be a designated shelter, school policy across the country is that if children cannot be returned home safely, they must be sheltered in place in the school until parents can pick them up. So even if a school is not officially designated as a shelter, school policies have made them into de facto shelters.

The 1933 Long Beach magnitude-6.4 earthquake in southern California is best known for damaging thousands of URM buildings, including over 230 school buildings. Fortunately, school had ended for the day at the time of the earthquake. Had that not been the case, thousands of children would have been injured or killed. The outcry from seeing collapsed school buildings directly led to the State of California passing the Field Act, which mandated earthquake-resistant construction requirements and inspection for all future school buildings.

While the January 1994 magnitude-6.7 Northridge earthquake in southern California did not collapse any school buildings, the amount of damage, including collapsed suspended ceilings and light fixtures, would have injured children had the earthquake occurred during school hours. Fortunately, the earthquake occurred early in the morning on a national holiday. Even so, FEMA funded a major seismic retrofitting program to seismically brace all suspended ceilings and light fixtures in every Los Angeles County school building.

While there have been notable efforts by some states, particularly Oregon and Utah, to identify at-risk school buildings and to begin the process of addressing the seismic risk they present, they have all been severely limited by budget issues and the day-to-day problems local governments face to just to keep their schools operating.

However, this is not just an earthquake problem. In May 2013, an EF5 tornado struck Moore, Oklahoma and resulted in 24 fatalities, including seven children at the Plaza Towers Elementary School. In April 2014, an EF4 tornado leveled a brand-new school still under construction in Vilonia, a suburb of Little Rock, Arkansas. While schools generally have some short-term notification of a tornado warning, and tornado safe rooms are becoming an accepted standard of care, and are now a requirement for new schools in certain locations under the *2015 International Building Code*, many schools remain vulnerable to tornadoes with no safe haven for students or staff.

The risk from flooding is generally well known and mapped, and sufficient warning time usually exists that the risk from this hazard is well controlled. However, the risk from flash flooding in mountainous terrain or from storm surge flooding in coastal areas can still be a significant hazard for schools located in harm's way. Severe flooding, as with other natural hazard events, can also lead to school closures and long-term negative impacts on students.

Despite the critical role that schools play in people's lives, many obstacles exist in attempting to improve school safety from natural hazards. These include competing public needs and demands, scarce resources in an increasingly difficult economic and political environment, and lack of understanding of the risk of natural hazards. We believe that a comprehensive document for school administrators and staff, as well as concerned parents that provides advice on both successful operational policies and practices, as well as recommendations on how to improve the physical protection of the school facility to resist applicable natural hazards would help improve overall school safety.

FEMA recently worked with the Department of Education and other federal partners to develop the *Guide on Developing High Quality School Emergency Operations Plans* (U.S. Department of Education, 2013), a school safety planning guide that covers a wide range of possible hazards and threats. The goal of FEMA P-1000 was to develop a companion guide that provides additional information specific to natural hazards to help schools be better prepared and better able to respond, recover, and mitigate future natural hazards. This *Guide* focuses on operational guidance (what to do before, during and after an event) as well as physical protection (what can be done to the structure and facility to improve safety). It was developed with input from design professionals, emergency managers, school administrators, teachers, representatives of concerned parent groups, and other relevant entities.

FEMA wishes to express its gratitude to the Project Management Committee (PMC) of Barry H. Welliver (Chair), Suzanne Frew, William T. Holmes, Christopher P. Jones, Lori Peek, John Schelling, Thomas L. Smith, and Edward Wolf. The PMC managed the development efforts and also served as principal authors. We also wish to thank Laura Dwelley-Samant, who was the Report Development Consultant, as well as Lucy Carter, Shawna Bendeck, Scott Kaiser, Jacob Moore, Meghan Mordy, Katherine Murphy, and Jennifer Tobin, who provided assistance in the literature search and focus group work.

FEMA also wishes to thank the Project Review Panel, which consisted of Ines Pearce (Chair), Jill Barnes, Victor Hellman, Andrew Kennedy (ATC Board Contact), Rebekah Paci-Green, and Cindy Swearingen. They provided review, advice, and consultation at key stages of the work. The names and affiliations of all who contributed to this report are provided in the list of Project Participants.

Without the dedication and hard work of all of these people, this project would not have been possible.

Federal Emergency Management Agency

Preface

In 2014, the Applied Technology Council (ATC), with funding from FEMA under Task Order Contract HSFE60-12-D-0242, commenced a two-year project (ATC-122) to develop a document that would provide school safety guidance to use before, during, and after a natural hazard event by updating existing documents and providing new information on improved knowledge about natural hazard-resistant design and policies and procedures recommended by other federal agencies. In particular, this project would build upon the *Guide for Developing High-Quality School Emergency Operations Plans* (U.S. Department of Education, 2013), which was developed as a multi-agency effort involving the Department of Education, the Department of Health and Human Services, the Department of Homeland Security and its Federal Emergency Management Agency, and the Department of Justice and its Federal Bureau of Investigation. To help inform the development of the document under the ATC-122 Project, the project team conducted a literature review of over 250 existing relevant resources and held videoconference calls with focus groups made up of representatives of the intended audience.

The resulting *Guide* provides up-to-date, authoritative information that schools can use to develop a comprehensive strategy for addressing natural hazards.

ATC is thankful for the leadership of Barry H. Welliver, Project Technical Director, and to the members of the ATC-122 Project Team for their efforts in developing this *Guide*. The Project Management Committee, consisting of Suzanne Frew, William T. Holmes, Christopher P. Jones, Lori Peek, John Schelling, Thomas L. Smith, and Edward Wolf, managed the development efforts and served as principal authors. Laura Dwelley-Samant served as the Report Development Consultant and Lucy Carter, Shawna Bendeck, Scott Kaiser, Jacob Moore, Meghan Mordy, Katherine Murphy, and Jennifer Tobin provided assistance in the literature search and focus group work as members of the Project Working Group. The Project Review Panel, consisting of Ines Pearce (Chair), Jill Barnes, Victor Hellman, Andrew Kennedy (ATC Board Contact), Rebekah Paci-Green, and Cindy Swearingen, provided review, advice, and consultation at key stages of the work. Focus group members, consisting of Debbie Carter-Bowhay, Cathy Coy, Susan Graves, Julie

Mahoney, Bob Roberts, Kerry Sachetta, Shawn Streeter, and Randy Trani, provided valuable feedback as representatives of the target audience. The names and affiliations of all who contributed to this report are provided in the list of Project Participants.

ATC is indebted to the leadership of Mike Mahoney (FEMA Project Officer) who conceived the project, contributed to development efforts, and provided guidance at critical stages. ATC also gratefully acknowledges Drew Herseth (FEMA Task Monitor) whose input and guidance made this document possible. ATC is thankful to John Westcott (FEMA) and Madeline Sullivan (U.S. Department of Education) for their review of this document. Veronica Cedillos and Ayse Hortacsu served as the ATC Project Managers and Carrie Perna provided report production services.

Ayse Hortacsu
ATC Director of Projects

Jon A. Heintz
ATC Executive Director

Table of Contents

Foreword iii

Preface vii

List of Figures xv

List of Tables xxi

Executive Summary xxiii

1. **An Introduction to School Natural Hazard Safety** 1-1
   1.1   Overview of Schools and Impacts of Natural Hazards 1-2
         1.1.1   Impacts on School Operations 1-2
         1.1.2   Vulnerabilities of School Buildings 1-3
         1.1.3   Exposure to Natural Hazard Events 1-3
   1.2   Purpose of this *Guide* 1-6
   1.3   Comprehensive Approach to Reducing Risk in Schools 1-6
         1.3.1   A Comprehensive Approach for Natural Hazards 1-8
         1.3.2   Goals of a Comprehensive Approach 1-9
   1.4   Requirements and Voluntary Measures for Schools 1-10
   1.5   How to Use this *Guide* 1-14

2. **Identifying Relevant Natural Hazards** 2-1
   2.1   An Overview of Natural Hazards 2-1
   2.2   Natural Hazards: Characteristics and Where They Occur 2-3
         2.2.1   Earthquakes 2-3
         2.2.2   Floods 2-6
         2.2.3   Hurricanes 2-7
         2.2.4   Tornadoes 2-9
         2.2.5   Tsunamis 2-10
         2.2.6   High Winds 2-12
         2.2.7   Other Hazards 2-13
   2.3   Summary Checklist – Which Hazards are Relevant to Your
         School? 2-13

3. **Making School Buildings Safer** 3-1
   3.1   School Building Safety from Natural Hazards 3-1
         3.1.1   Level of Safety Provided by Building Codes 3-3
   3.2   Existing School Buildings 3-4
         3.2.1   Determining Building Vulnerability 3-4
         3.2.2   Identifying and Evaluating Mitigation Options 3-6
         3.2.3   Developing an Implementation Plan 3-6
   3.3   New School Buildings 3-9
         3.3.1   Smart Site Selection 3-9
         3.3.2   Relevant Building Codes and Resilient Design 3-9

| | | 3.3.3 | Schools as Emergency Shelters or Recovery Centers | 3-10 |

3.3.3 Schools as Emergency Shelters or Recovery Centers 3-10
3.4 Developing a Funding Plan 3-11
3.5 Importance of Quality Assurance Measures 3-13
    3.5.1 Overview of Design and Construction 3-13
    3.5.2 Long-Term Maintenance and Improvements 3-13

4. Planning the Response **4-1**
4.1 Purpose of a School Emergency Operations Plan 4-2
4.2 Recommended Process to Develop an EOP 4-3
4.3 Overview of Structure and Content of an EOP 4-6
    4.3.1 The Basic Plan 4-6
    4.3.2 Functional Annexes 4-7
    4.3.3 Threat- or Hazard-Specific Annexes 4-9
4.4 Legislative Considerations in Developing EOPs 4-10
4.5 Training and Exercises 4-11
4.6 Making the Plan Actionable 4-12

5. Planning the Recovery **5-1**
5.1 Getting Back in School Buildings 5-1
    5.1.1 Post-Disaster Building Assessment 5-1
    5.1.2 Documenting the Damage 5-2
    5.1.3 Building Back Better 5-2
    5.1.4 Adaptability 5-4
    5.1.5 Schools as Emergency or Recovery Shelters 5-4
5.2 Focusing on Routine and Mental Health 5-6
    5.2.1 Re-Establishing Routine 5-6
    5.2.2 Assessing Mental Health 5-7
5.3 Financing the Recovery 5-8
    5.3.1 Federal Resources 5-9
5.4 Planning for the Next One 5-11

6. Engaging the Whole Community **6-1**
6.1 The Whole Community Approach 6-1
6.2 Engage Community Partners 6-3
    6.2.1 Children and Youth 6-4
    6.2.2 Design Professionals 6-7
    6.2.3 Educational Professionals 6-7
    6.2.4 Elected Officials 6-7
    6.2.5 Emergency Management Professionals 6-7
    6.2.6 Labor Bargaining Units 6-8
    6.2.7 Local Business and Industry 6-8
    6.2.8 Local Community Organizations 6-8
    6.2.9 Local Hospitals 6-9
    6.2.10 Local Jurisdiction Public Agencies 6-9
    6.2.11 Media 6-9
    6.2.12 Parents and Caregivers 6-10
6.3 Communicating with the Community 6-11
    6.3.1 Before the Event 6-11
    6.3.2 During the Emergency and Recovery Phase 6-13
    6.3.3 After the Recovery Phase 6-13
6.4 Tools and Technology for Effective Communication 6-14

7. **Moving Forward**...**7-1**
    7.1   Identified Challenges..7-1
    7.2   Potential Opportunities...7-3

Supplement E: Earthquakes **E-1**
    E.1   Overview of Earthquakes E-1
          E.1.1   Earthquake Impacts on Schools E-2
    E.2   Is Your School in an Earthquake-Prone Region?....................E-4
          E.2.1   Determining the Severity of the Hazard....................E-5
          E.2.2   Determining Your School's Vulnerability E-7
    E.3   Making Buildings Safer...E-8
          E.3.1   Existing School Buildings.......................................E-8
          E.3.2   New School Buildings...E-14
          E.3.3   Nonstructural Systems and Contents......................E-15
    E.4   Planning the Response..E-19
          E.4.1   During the Earthquake...E-20
          E.4.2   Immediately Following Shaking E-21
    E.5   Planning the Recovery...E-22
    E.6   Recommended Resources..E-23

Supplement F: Floods **F-1**
    F.1   Overview of Floods F-1
          F.1.1   Flood Impacts on Schools F-2
    F.2   Is Your School in a Flood-Prone Region?..............................F-3
          F.2.1   Flood Hazard Maps F-4
          F.2.2   Levees and Other Flood Control Structures F-6
    F.3   Making Buildings Safer...F-6
          F.3.1   Existing School Buildings.......................................F-8
          F.3.2   New School Buildings...F-12
    F.4   Planning the Response..F-14
    F.5   Planning the Recovery...F-15
    F.6   Recommended Resources..F-16

Supplement H: Hurricanes **H-1**
    H.1   Overview of Hurricanes H-1
          H.1.1   Hurricane Impacts on Schools................................H-3
          H.1.2   Improvements in School Construction for
                   Hurricanes H-7
          H.1.3   Important Terminology H-8
    H.2   Is Your School in a Hurricane Hazard Area? H-8
    H.3   Making Buildings Safer...H-11
          H.3.1   Existing School Buildings.......................................H-11
          H.3.2   New School Buildings...H-15
    H.4   Planning the Response..H-17
    H.5   Planning the Recovery...H-18
    H.6   Recommended Resources..H-20

Supplement TO: Tornadoes..**TO-1**
    TO.1 Overview of Tornadoes TO-1
          TO.1.1 Tornado Impacts on Schools TO-2
          TO.1.2 Important Terminology TO-5
    TO.2 Is Your School in a Tornado-Prone Region? TO-5

FEMA P-1000   **Table of Contents**   **xi**

TO.3 Protecting Occupants within School Buildings TO-7
    TO.3.1 Existing School Buildings TO-7
    TO.3.2 New School Buildings TO-10
TO.4 Planning the Response TO-13
TO.5 Planning the Recovery TO-15
TO.6 Recommended Resources TO-16

Supplement TS: Tsunamis...TS-1
TS.1 Overview of Tsunamis...TS-1
    TS.1.1 Causes of Tsunamis TS-2
    TS.1.2 Local or Distant?..TS-3
TS.2 Is Your School in a Tsunami Hazard Zone?.....................TS-5
    TS.2.1 Mapped Tsunami Hazard Zones...................TS-6
    TS.2.2 Unmapped Tsunami Hazard Zones TS-6
TS.3 Tsunamis and School Buildings TS-7
    TS.3.1 General Considerations..............................TS-7
    TS.3.2 Existing School Facilities TS-8
    TS.3.3 New Facilities: Design for Vertical Evacuation TS-8
    TS.3.4 Schools as Evacuation Shelters or Refuges TS-10
TS.4 Planning the Response...TS-10
    TS.4.1 Warning Signs TS-11
    TS.4.2 Plans, Policies, Procedures TS-12
    TS.4.3 Evacuation Protocols and Practice...............TS-15
TS.5 Planning the Recovery TS-19
TS.6 Recommended Resources TS-19

Supplement W: High Winds..W-1
W.1 Overview of High Winds..W-1
W.2 Is Your School in a Region Exposed to High Winds?...........W-2
W.3 Making Buildings Safer...W-2
    W.3.1 Existing School Buildings W-2
    W.3.2 New School Buildings W-3
W.4 Planning the Response W-4
W.5 Planning the Recovery W-4
W.6 Recommended Resources W-5

Supplement X: Other Hazards...X-1
X.1 Snow Storms..X-1
X.2 Volcanic Eruptions X-2
X.3 Wildfires X-3
X.4 Recommended Resources X-3

Appendix AE: Earthquake Appendix AE-1
AE.1 Determination of Seismicity Region of Site AE-1
AE.2 Adequacy of Building Codes....................................AE-4

Appendix AF: Flood Maps Appendix................................AF-1
AF.1 Understanding and Using Flood Hazard Maps.....................AF-1
AF.2 FIRMs and Related Products AF-2
    AF.2.1 Finding FIRMs and Related Products.....................AF-3
    AF.2.2 Reading FIRMs......................................AF-5
AF.3 Other Types of Flood Hazard Maps AF-8

AF.3.1 Coastal Flood Hazard Map Comparison AF-8
AF.3.2 Riverine Flood Hazard Map Comparison............... AF-10
AF.3.3 Future Conditions Affecting Flood Hazards AF-11

Appendix AR: Resources **AR-1**

AR.1 Communications..AR-1

AR.2 Community Engagement.....................................AR-2

AR.3 Curriculum AR-2

AR.4 Disaster and Emergency Planning........................AR-4

AR.4.1 General Guidance.....................................AR-4

AR.4.2 Federal Laws Applicable to Emergency
Operations Plans......................................AR-7

AR.4.3 Incident Command System AR-8

AR.4.4 School Safety Plan Examples...................AR-8

AR.5 Emergency Exercises, Drills, and Materials..........AR-9

AR.6 International Resources AR-9

AR.7 Mental Health...AR-10

AR.8 Vulnerability Assessments AR-12

References.. **R-1**

Project Participants **P-1**

List of Figures

Figure 1-1   Types of state emergency management resources shown on the interactive map on the REMS website....... 1-12

Figure 1-2   2015 disaster report card... 1-13

Figure 2-1   Relative seismic hazard map showing earthquake intensity.. 2-4

Figure 2-2   Map indicating hurricane hazard areas in the United States.. 2-8

Figure 2-3   Map indicating tornado-prone regions in the United States.. 2-10

Figure 3-1   Tornado safe room placard 3-8

Figure 3-2   Thanks to a federal grant, a new monolithic dome resembling the existing Beggs Event Center will house Beggs School District's tornado safe room 3-12

Figure 4-1   Six step process to develop, review, approve, and maintain a school Emergency Operations Plan.................. 4-4

Figure 4-2   Example Incident Command System chart........................ 4-5

Figure 4-3   Example structure of a school Emergency Operations Plan using a traditional format.. 4-6

Figure 4-4   Participants taking part in the tabletop exercise............... 4-13

Figure 5-1   Photos documenting earthquake damage to school buildings... 5-3

Figure 5-2   Schoolchildren affected by the Colorado floods................ 5-5

Figure 5-3   Drawing by child affected by Hurricane Katrina............... 5-7

Figure 5-4   Briarwood Elementary was one of three public schools destroyed or damaged by the Moore, Oklahoma tornado of May 2013.. 5-10

Figure 6-1   Young people in Joplin, Missouri shared their stories through the YCDR^2 project... 6-5

Figure 6-2   A member of Teen CERT explains why the experience has been so valuable to him 6-6

Figure 7-1	The three pillars of the Comprehensive School Safety Framework	7-2
Figure 7-2	Map indicating the percentage of low income students in U.S. public schools	7-4
Figure E-1	Site of a student fatality in the 1949 Olympia earthquake in Washington	E-2
Figure E-2	Examples of nonstructural damage	E-4
Figure E-3	Relative seismic hazard map of the United States	E-6
Figure E-4	Collapse of part of Jefferson High School in the 1933 Long Beach earthquake	E-10
Figure E-5	Collapse of a portion of a concrete school in the Helena, Montana earthquake of 1935	E-11
Figure E-6	High school first floor plan indicating the identified potential spaces for shelter sleeping and other important planning considerations for shelter operations	E-16
Figure E-7	Fallen light fixtures in a classroom after the Northridge earthquake	E-17
Figure F-1	Ramstad Middle School, in Minot, North Dakota was flooded in June 2011, badly damaged, and ultimately decommissioned in June 2012	F-2
Figure F-2	Graphic illustrating the rise in nuisance flooding around the United States, but especially off the East Coast	F-4
Figure F-3	This school in Puerto Rico was undermined by riverine erosion during Hurricane Georges in September 1998	F-5
Figure F-4	School buses in New Orleans, Louisiana were swamped by the floodwaters following Hurricane Katrina in September 2005	F-7
Figure F-5	Danville Middle School in Pennsylvania flooded in September 2011	F-10
Figure F-6	Wall repair and reconstruction to resist future flood damage	F-11
Figure F-7	Metal lockers were replaced with high density polyethylene (HDPE) lockers	F-11
Figure F-8	Flood vents were installed throughout most of the school	F-11

Figure H-1	Storm surge rises above the normal tide	H-2
Figure H-2	This book documented the many long-term, negative effects that Hurricane Katrina had on children	H-3
Figure H-3	Large portions of the school roof coverings blew off during 2005 Hurricane Katrina	H-4
Figure H-4	Damage to school ceiling in Mississippi during 2005 Hurricane Katrina	H-4
Figure H-5	Storm surge flooding in a high school in LaPlace, Louisiana during Hurricane Isaac in 2012	H-5
Figure H-6	Storm surge flooding of a high school in Sabine Pass, Texas during Hurricane Ike in 2008	H-5
Figure H-7	Denham Springs, Louisiana, High School flooded in August 2016	H-6
Figure H-8	Timeline indicating a selection of significant hurricane events and improvements to building codes and guidelines	H-7
Figure H-9	Recorded Category 1–5 hurricanes striking the Atlantic and Gulf coastal areas and Hawaii from 1950 to 2014	H-9
Figure H-10	Map indicating hurricane hazard areas in the United States	H-9
Figure H-11	Example hurricane storm surge inundation map in the Norfolk, Virginia area	H-10
Figure H-12	Damage in a school in Florida during Hurricane Andrew in 1992	H-11
Figure H-13	Collapsed unreinforced masonry classroom wall in the U.S. Virgin Islands during Hurricane Marilyn in 1995	H-12
Figure H-14	Rooftop mechanical equipment is often inadequately attached	H-12
Figure H-15	This school was retrofitted with accordion shutters prior to a hurricane	H-14
Figure H-16	Had this tree fallen in the opposite direction during 2004 Hurricane Ivan in Florida, it would have damaged the school	H-14
Figure H-17	View of aid tents between a damaged elementary school and high school in Florida after Hurricane Andrew	H-17

FEMA P-xxx   **List of Figures**   **xvii**

Figure H-18	Crenshaw Elementary and Middle School, located in Port Bolivar, Texas, which was successfully designed to be a hurricane shelter in 2005 and was one of the few buildings standing after 2008 Hurricane Ike	H-18
Figure H-19	Essentially the entire roof covering on this older school blew off	H-19
Figure TO-1	Enhanced Fujita scale	TO-2
Figure TO-2	Typical tornado damage descriptions particular to schools and their corresponding intensity according to the EF Scale.	TO-3
Figure TO-3	Aerial view of a school devastated by a strong tornado in Moore, Oklahoma in 2013	TO-4
Figure TO-4	Wind-borne debris in a classroom in Greensburg, Kansas in 2007	TO-4
Figure TO-5	Recorded EF0 and EF1 tornadoes from 1950 to 2014	TO-6
Figure TO-6	Recorded EF2, EF3, EF4, and EF5 tornadoes from 1950 to 2014	TO-6
Figure TO-7	Safe room/shelter design wind speed zones for tornadoes	TO-7
Figure TO-8	This multipurpose room addition was designed as a safe room in Wichita, Kansas	TO-8
Figure TO-9	Identified refuge areas, both previous and updated, marked on school floor plan	TO-9
Figure TO-10	Timeline of development and requirements of tornado safe rooms/shelters	TO-11
Figure TO-11	View of the gymnasium	TO-12
Figure TO-12	Post-tornado view	TO-14
Figure TO-13	Damaged hallway where seven fatalities occurred	TO-14
Figure TS-1	Tsunami sources	TS-2
Figure TS-2	Impacts of the 1964 Alaska tsunami on lower Ecola Creek, Cannon Beach, Oregon	TS-4
Figure TS-3	Tsunami evacuation map for Crescent City, California	TS-7

Figure TS-4	Ocosta Elementary School's heavily reinforced vertical evacuation structure under construction in Westport, Washington	TS-9
Figure TS-5	Tsunami evacuation route sign	TS-16
Figure TS-6	Waves surge past Togura Elementary School in Minami-Sanriku, Japan during the 2011 Great East Japan Tsunami	TS-18
Figure W-1	The roof on this new school blew off during moderate winds soon after it was installed	W-2
Figure W-2	The roof on this school blew off during cold weather, thus increasing the repair costs	W-3
Figure AE-1	Screenshot from USGS seismic design parameter calculator	AE-2
Figure AE-2	Screenshot showing parameters to determine Seismicity Region	AE-3
Figure AF-1	Screenshot from FEMA MSC website with the street address of Tanglewood Elementary School entered	AF-4
Figure AF-2	Screenshot showing MSC products available for Tanglewood Elementary School	AF-4
Figure AF-3	Portion of screenshot of NFHL image of Tanglewood Elementary School	AF-5
Figure AF-4	Generalized depiction of a FEMA Flood Insurance Rate Map (FIRM), showing flood hazard zones and Base Flood Elevations	AF-7
Figure AF-5	Portion of screenshot of the portion of the National Flood Hazard Layer for Tampa, Florida	AF-8
Figure AF-6	Storm surge atlas for Tampa, Florida	AF-9
Figure AF-7	Portion of screenshot from the National Flood Hazard Layer for Lawton, Oklahoma	AF-10
Figure AF-8	Dam failure inundation map for the area shown in Figure AF-7	AF-11
Figure AF-9	Charlotte-Mecklenburg, North Carolina, flood hazard map showing 1% annual chance flood hazard area and community flood hazard area based on future development upstream	AF-12
Figure AF-10	Areas that potentially could be impacted by the 100-yr flood and various sea level rise heights	AF-13

Figure AF-11   Increased flooding scenario brochure for San Francisco Bay Area AF-14

List of Tables

Table A	Organization of this *Guide*	xxiv
Table 1-1	Organization of this *Guide*	1-14
Table 2-1	Natural Hazards and Other Types of Hazards and Threats	2-2
Table 2-2	Coastal Areas in the United States Ranked by Tsunami Hazard	2-11
Table 3-1	Likely Warning Times by Hazard and Resulting School Occupancy Expectation During Hazard Events	3-2
Table E-1	Times of Damaging Earthquakes in the United States Since 1906	E-3
Table TS-1	Coastal Areas in the United States Ranked by Tsunami Hazard	TS-5
Table AE-1	Determination of Seismic Region from USGS Design Parameters, S_s and S_1	AE-4
Table AE-2	Building Codes Presumed to be Adequate (if Properly Followed) by FEMA P-154	AE-5
Table AF-1	List of Flood Hazard Zones that May Be Shown on a FIRM	AF-6

Executive Summary

Will we value and invest in school disaster resilience for the sake of our children's safety and the future of our communities? Or will we fail to act until after our schools and communities experience irrecoverable loss that could have been prevented? This is a choice and we choose the former. We hope you do the same. —anonymous

> The stakes are high—natural hazards can endanger the lives of children and staff, increase emotional traumas, and result in long-term harm to children and communities.

School leaders and state officials are the specialists and authorities for educating and protecting our children. But natural hazards may be unfamiliar territory, requiring skills, plans, and support to which school communities may not have access. Poor building performance during a disaster is exacerbated by inadequate strategies to prepare for, respond to, recover from, and mitigate against natural disasters. In contrast, schools that have taken steps to reduce their risks and have adequately prepared for emergencies can respond effectively, recovery quickly, and help support the entire community to recover from a disaster.

This *Guide* provides up-to-date, authoritative information and guidance that schools can use to develop a comprehensive strategy for addressing natural hazards. The *Guide* presents information and guidance on:

- Identifying natural hazards that could potentially impact a school;

- Making new and existing school buildings safer for children and staff, and more resistant to damage during natural disasters;

- Planning and preparing for effective and successful response during a natural disaster;

- Recovering after a natural disaster as quickly and robustly as possible, and being better prepared for future events; and

- Engaging the whole community in the entire process in order to improve school and community disaster resilience.

This *Guide* is intended to be used by administrators, facilities managers, emergency managers, emergency planning committees, and teachers and staff at K through 12 schools. It can also be valuable for state officials, district administrators, school boards, teacher union leaders, and others that play a role in providing safe and disaster-resistant schools for all. Parents,

FEMA P-1000   **Executive Summary**   **xxiii**

caregivers, and students can also use this *Guide* to learn about ways to advocate for safe schools in their communities.

This *Guide* is divided into three main sections, as shown in Table A. This *Guide* only focuses on natural hazards. It does not cover other types of important hazards and threats, including technological or intentional threats, as there are resources already available to schools that address these. In particular, this *Guide* builds upon the *Guide for Developing High-Quality School Emergency Operations Plans* (U.S. Department of Education, 2013), which provides the latest guidance on developing school emergency operation plans that is applicable to many hazards and threats.

School leadership and state officials are busy with their critical work to educate and protect children. It is a challenge to focus limited resources and time on planning and preparing for rare and complicated events like natural disasters. But doing so is essential. It can mean the difference between life and death, or the difference between a devastated community and one that recovers quickly and effectively, becoming even stronger than before. It can also make all the difference in the life and educational trajectory of a child. Ultimately, this *Guide* provides actionable guidance to help school leaders take necessary steps to be as ready as possible when the next disaster strikes.

Table A **Organization of this *Guide***

Section	Content	Description
Comprehensive Approach for School Natural Hazard Safety	Ch 1: An Introduction to School Natural Hazard Safety Ch 2: Identifying Relevant Hazards Ch 3: Making School Buildings Safer Ch 4: Planning the Response Ch 5: Planning the Recovery Ch 6: Engaging the Whole Community Ch 7: Moving Forward	These chapters are recommended for all readers. They provide an overview of the key components of a comprehensive approach for school natural hazard safety.
Hazard-Specific Supplements	E: Earthquakes F: Floods H: Hurricanes TO: Tornadoes TS: Tsunamis W: High Winds X: Other Hazards: Snow Storms, Volcanic Eruptions, and Wildfires	These supplements provide detailed information and guidance focused on particular natural hazards. Readers should refer to the hazards that affect their schools.
Appendices	Earthquake Appendix Flood Maps Appendix Resources Appendix	The appendices provide more detailed information on a variety of topics, and are referenced in other sections of this *Guide*. The *Resources Appendix* points the reader to many sources of useful information that expand on the topics covered by this *Guide*.

Chapter 1

An Introduction to School Natural Hazard Safety

Schools are fundamentally important places. Schools—and here we are referring to the buildings and, importantly, to the adults and children in those structures on any given school day—are where our future generations are educated.

> 94% of American children live in communities at risk of natural disasters (Save the Children, 2012).

Schools are also unique in terms of the risks they face and responsibilities they have associated with natural disasters. Because natural hazard events are infrequent, they can be seen as less urgent and are more easily demoted in importance when compared to other more urgent, daily challenges faced by school leaders. But school administrators and state officials have a moral, and in many cases legal, responsibility to make their schools more resilient to disasters and to minimize the risk of damage and injury in natural hazard events. Properly planning for natural hazards results in safer, more prepared and resilient schools and the benefits from ensuring school safety go beyond the schoolyard—schools can help drive the health, prosperity, and quality of life for an entire community.

This chapter provides an introduction to school natural hazard preparedness and safety by covering the following:

- An overview of how natural hazards impact schools, including issues that school administrators and local and state officials need to consider to address risks from natural hazards;

- The purpose of this *Guide* and how it relates to the recently published *Guide for Developing High-Quality School Emergency Operations Plans* (U.S. Department of Education, 2013);

- A description of key elements of a comprehensive approach to mitigate the effects of natural hazards, to protect the school community, and to effectively respond and recover from potential events, including earthquakes, floods, tsunamis, tornadoes, hurricanes, and other natural hazards;

- A description of relevant requirements and voluntary measures related to natural hazard safety that are applicable to schools; and

---

FEMA P-1000   1: An Introduction to School Natural Hazard Safety   1-1

- A summary of how this *Guide* is structured and how to use it.

1.1 Overview of Schools and Impacts of Natural Hazards

> It took 7 months for the last Katrina child to be reunited with a parent (Save the Children, 2015).

Natural disasters can have both immediate and long-term impacts on school buildings, their occupants, and the surrounding community. Most devastatingly, natural disasters can cause deaths and injuries among students and staff when structures are unsafe or located in areas vulnerable to natural hazards.

Children depend on safe school buildings and school staff to ensure their safety and well-being both during and immediately after a disaster that occurs during school hours. Due to the disruption and damage that can accompany a disaster event, school officials, staff, and teachers may need to provide care for some children for an extended period of time before they are reunited with parents or guardians. Experiences during a disaster and the long-term recovery period that follows can result in emotional trauma to both students and staff.

When students cannot attend school, the entire community and its ability to recover is affected. Not only is students' education interrupted, their routine disrupted, and the school services they receive suspended, but parents of younger children often cannot return to work or volunteer, thus stalling the recovery process for parents and children alike. It is becoming increasingly clear that the recovery of the entire community is linked to the resilience of schools.

1.1.1 Impacts on School Operations

> Disaster impacts on school operations can include: facility damage, staff shortages, and the use of the school building as either an evacuation or community housing/recovery shelter. All potential impacts should be considered in the school emergency operations plans.

The days and weeks following disaster events can severely impact school operations. Damage to individual school buildings, school sites, roads, and utilities can lead to school closures that last weeks, months, or even over a year. School operations may also be affected by lack of staff, which may be a result of delayed returns from large-scale evacuations or staff living in severely impacted neighborhoods. School buildings are also often designated as shelters to be used by the community following a disaster. Schools are often selected as shelters because of their prime locations in the community and their availability of large spaces, such as gymnasiums and auditoriums. Although this can be very helpful for communities following disasters, it can also impede school operations. These impacts on school operations—whether they are caused by physical damage to buildings and infrastructure, staff shortages, or the use of a school building as a shelter— should all be considered in school emergency operations plans (EOPs).

1-2   **1: An Introduction to School Natural Hazard Safety**   **FEMA P-1000**

1.1.2 Vulnerabilities of School Buildings

In the United States, school buildings are the only high occupancy public buildings, other than prisons and courthouses, whose inhabitants are compelled by legal mandate to be inside them. Perhaps because children are required to attend school, by law, the general public often perceives school buildings as possessing good resistance to natural hazards. However, school buildings can be more vulnerable to damage in natural hazards than other types of buildings. For example, schools frequently have large assembly rooms, such as gymnasiums and auditoriums, which are more vulnerable to damage in natural hazards such as earthquakes, hurricanes, and tornadoes. Furthermore, school buildings often remain in use for many more years than other types of buildings and may not receive consistent capital renewal funds. Although building codes change and improve continually as building professionals learn more about how to design disaster-resistant buildings, many school buildings are decades old and, thus, were constructed to older building code standards. Meaning, older school buildings are particularly vulnerable to natural hazards and in most cases, school administrators do not have the financial resources to address these vulnerabilities.

In many cases, schools are pre-designated as emergency shelters, yet they have not been designed or constructed to the standards that will ensure that they will even be occupiable following a disaster. Both the designation and operation of shelters in school buildings take careful consideration and planning, from building code and performance requirements to operational considerations. The core planning team, which should include local emergency management organizations, should have pre-established agreements in place as part of the school EOP.

> During normal working hours—which total more than 2,000 hours a year—the safety of nearly 68 million of our country's children is in the hands of school officials and child caregivers. Most parents assume that when they drop their kids off for the day, they will be safe if disaster strikes. But two-thirds of U.S. states have not adopted basic emergency preparedness regulations for child care facilities and schools (Save the Children, 2012).

> The core planning team per the *Guide for Developing High-Quality School Emergency Operations Plans* should include a wide range of people, including representatives of school personnel, students, parents, and community partners.

1.1.3 Exposure to Natural Hazard Events

Over the last several decades, the United States has experienced an escalation in the number of damaging natural hazard events and their corresponding costs resulting from that damage. Despite this increase, loss of life in schools due to natural disasters has been reduced over time. This is due to several factors, including more stringent building codes for newly constructed schools, better state regulations regarding school building inspections and construction, the inclusion of professional emergency managers in school preparedness, mandated emergency planning, and better warning systems.

The increase in disaster frequency is not only felt in the United States, but also around the world. For example, earthquakes in recent years have caused massive destruction of schools and high number of fatalities. In 2005, an earthquake in Pakistan caused 6,700 school buildings to collapse during school hours, killing approximately 17,000 students. In 2008, in Sichuan, China, at least 7,000 children died in the collapse of school buildings. Although earthquake disasters of similar magnitude have not hit the United States during school hours in contemporary times, the potential is still very real and should serve as a warning.

A Tale of Two Earthquakes

Two earthquakes occurred just ten months apart in the late 1980s. Both were magnitude-6.9 earthquakes. Both hit heavily populated urban areas. Both affected zones with housing, schools, and other critical infrastructure. But while these two earthquakes share many similarities, the devastation to school buildings and children could not be more different.

On December 7, 1988, an earthquake struck the country of Armenia during school hours at 11:41 am. Unfortunately, buildings in the affected area had been built with little to no consideration of earthquake effects. Half a million buildings were destroyed including over 900 schools. About 25,000 people died in the destruction and it is estimated that 6,000 of the deaths were school children (Balassanian et al., 1995).

On October 17, 1989, an earthquake of the same magnitude hit Loma Prieta, California. In contrast to the Armenian earthquake, the Loma Prieta earthquake resulted in only 63 deaths. And while this earthquake took place after school hours, no school buildings collapsed, and only three schools were severely damaged. Luckily for the affected area, the 1933 Field Act—which was enacted in response to the 1933 Long Beach earthquake that destroyed hundreds of schools, fortunately after school hours—ensured the construction of safe school buildings able to withstand the earthquake of 1989.

Comparing these two cases, it is clear that the implementation of strict building codes and mitigation measures in California law and practices saved many lives. While schools in California are held to higher standards in terms of earthquake resistance, many schools around the United States are not and are still at risk. Are we willing to accept this risk?

School Buildings get a D+

Every few years, the American Society of Civil Engineers (ASCE) issues Infrastructure Report Cards that indicate the current conditions of infrastructure. The 2017 report card gave school infrastructure a D+, noting that the nation continues to underinvest in school facilities and that more than 53% of public schools need to make investments to be in "good" condition (ASCE, 2017a).

The public may expect or assume that their safety from natural hazards is assured, but the reality is that, as of 2012, the average age of public school buildings was 44 years old (Alexander and Lewis, 2014). While some major renovations may have taken place in the interim, the original construction of numerous school buildings predates many of the modern building code requirements protecting occupants from natural hazards such as earthquakes, floods, high winds, and tsunamis and should raise concern. In fact, most older school buildings are of unreinforced masonry construction, which is arguably the most vulnerable type of building to earthquake and wind damage. Perhaps the most important actions for state and school officials to take include: (1) addressing the vulnerability of the large number of existing school facilities to natural hazards; (2) incorporating best practices for the design and construction of new or replacement school buildings; and (3)

There have been many "close call" events that would have caused extensive loss of life had the disaster occurred during the school day. Even so, many of these close calls have caused significant building damage and economic hardships, and have been disruptive to schools and the education process. Significantly reducing these risks and improving school building safety will require a concerted effort of all risk bearers, including state officials, school administrators, and entire communities.

Close Call in Joplin, Missouri

Imagine the following scenario. A tornado touches down in the heart of your city. The EF-5 tornado blows through town, causing $2.8 billion in damage and taking 161 lives. About 7,500 students attend school in your district, with over 2,000 of those students at the local high school. The schools in your district do not have storm shelters. The tornado flattens your high school along with a junior high and three elementary schools. How many lives would be lost if this disaster struck during an average school day? Thankfully, the town of Joplin, Missouri will never truly know, as these events occurred on Sunday, May 22, 2011. Had this EF-5 tornado struck on a school day, the results would have been more catastrophic.

Close Call in West Virginia

Recent flash floods in West Virginia were another close call for our schools. On June 23, 2016, heavy rainfall caused rivers to overflow and city streets, homes, and buildings, including schools, were suddenly flooded, floor to ceiling, with water, mud, and hazardous materials standing for days. Over two dozen schools went under water. While this flooding took place during summer break, it is difficult to imagine the disruption that may have occurred had this event taken place with schools in session. West Virginia was lucky. How many close calls are we willing to accept when it means putting the integrity of our school buildings and the safety of our children at risk?

Close Call in Coalinga, California

Imagine another scenario: this time, a magnitude-6.5 earthquake rocks your town. At the time the earthquake strikes, there are 1,900 students in attendance at the five schools throughout the city. While the school buildings in your district are built to have structural frames and walls withstand an earthquake of this intensity, other portions of the building, referred to as the nonstructural components of the buildings, have been ignored. Glass windows implode spraying glass everywhere within the schools. Thousands of light bulbs, fixtures, and ceiling tiles fall on the children. Water pipes burst and children are trapped in flooding classrooms. Sulfuric acid from the chemistry lab spills and eats through the flooring and lands on students below. Toxic fumes fill the classrooms and hallways, while everything from cabinets, book cases, typewriters, and television screens fly through the air.

How many fatalities would result from this earthquake? How many injuries to students, teachers, and staff might occur? This event actually took place in the City of Coalinga, California on May 2, 1983 with just one minor difference. The Coalinga earthquake struck just after schools had closed their doors for the day.

FEMA P-1000   1: An Introduction to School Natural Hazard Safety   1-5

> As of May 2017, FEMA has awarded over $660 million in federal funds for K-12 schools through their Hazard Mitigation Assistance programs. For more information on the grant programs and eligibility, refer to the current guidance for Hazard Mitigation Assistance at www.fema.gov/hazard-mitigation-assistance.

collaborating and coordinating with community partners to develop a comprehensive school EOP that supports the school community before, during, and after disasters. Although these actions can be challenging to implement given scarce resources and other competing needs and demands, there are opportunities and resources that can provide the necessary support, many which are highlighted in this *Guide*.

1.2 Purpose of this *Guide*

The purpose of this *Guide* is to provide the latest information and guidance on developing effective strategies for reducing risk in schools from natural hazards. It is intended to be used by administrators, facilities managers, emergency managers, emergency operations core planning team, and teachers and staff at K through 12 schools. It can also be valuable for state officials, district administrators, school boards, teacher union leaders, community partners, and others that play a role in providing safe and disaster-resistant schools for all. Parents, caregivers, and students can also use this *Guide* to learn about ways to advocate for safe schools in their communities. Finally, this *Guide* can also be used by all audiences to help communicate and promote the importance of school natural hazard safety.

This *Guide* builds upon the *Guide for Developing High-Quality School Emergency Operations Plans* (U.S. Department of Education, 2013). That document is referred to as the *School EOP Guide* throughout this *Guide*. The *School EOP Guide* was developed as a multi-agency effort, involving the Department of Education, the Department of Health and Human Services, the Department of Homeland Security and its Federal Emergency Management Agency, and the Department of Justice and its Federal Bureau of Investigation. The *School EOP Guide* provides the latest guidance on developing school emergency operation plans (EOPs) that is applicable to many hazards.

This *Guide* highlights and expands on information from the *School EOP Guide* that is particularly important for natural hazards. This *Guide* also provides detailed information and recommendations on natural hazards that are critical to the safety of school children, staff, and visitors.

1.3 Comprehensive Approach to Reducing Risk in Schools

A comprehensive approach for school natural hazard safety must incorporate many different actions for reducing risk. One framework that is particularly valuable for developing a comprehensive approach is outlined in the *School EOP Guide*. This particular resource is structured in terms of courses of action taken before a natural hazard event, during the hazard emergency, and after the emergency has ended. School emergency management and preparedness practitioners work to build capacity in the following five mission areas (adapted from the *School EOP Guide*).

> Risk from natural hazards is a combination of the severity of the hazard (e.g., location and intensity of an earthquake) and the vulnerability of the asset or institution under consideration (e.g., vulnerability of a school building).

Before the Hazard Event

- **Prevention** refers to actions to avoid or deter an incident from occurring. Of course, it is not possible to prevent most natural hazards—such as an earthquake, hurricane, tornado, tsunami, or volcanic eruption—from occurring. However, there are some available preventative measures, such as controlling adjacent vegetation to prevent wildfires from impacting a school.

- **Protection** refers to actions to secure buildings against natural disasters. This focuses on the ongoing actions that protect students, teachers, staff, visitors, and property from a hazard. Natural hazard-specific examples include practicing safety drills and developing policies and guidance for ongoing site-based assessments and disaster planning.

- **Mitigation** refers to actions to eliminate or reduce the loss of life and property damage by lessening the impact of an event or emergency. Natural hazard-specific examples include bracing or strapping file cabinets and bookshelves in earthquake-prone schools and seismically retrofitting school buildings to reduce damage.

During the Hazard Emergency

- **Response** refers to actions to stabilize an emergency once it has already happened or is certain to happen in an unpreventable way; to establish a safe and secure environment; to save lives and property; and to facilitate the transition to recovery. Natural hazard-specific examples include responding by engaging in "Drop, Cover, and Hold On" during earthquake shaking, evacuating to either high ground or a vertical evacuation refuge during a tsunami warning, or seeking shelter in a tornado safe room during a tornado warning. This would also include sheltering students in place during a natural hazard event until parents can safely pick up their children.

After the Event

- **Recovery** refers to actions to restore the learning environment for schools affected by an event. Recovery is an extended period that blends into the "before" timeframe of the next hazard event for a community, and should include steps to build back better so that future natural hazards have lesser impacts.

FEMA P-1000   1: An Introduction to School Natural Hazard Safety   1-7

1.3.1 A Comprehensive Approach for Natural Hazards

Building on the current guidance from the *School EOP Guide*, this *Guide* uses a structure that is well-suited for considerations particular to natural hazard safety in schools. In particular, the *Guide* is organized as follows:

- **Identifying Relevant Natural Hazards (Chapter 2).** Identifying and prioritizing which hazards could potentially impact a school is crucial to effectively planning and preparing for natural hazards. This chapter provides up-to-date resources and general information on how to determine whether a school is at risk from earthquakes, floods, hurricanes, tornadoes, tsunamis, high winds, and other natural hazards.

- **Making School Buildings Safer (Chapter 3).** Properly designed, constructed, and maintained school buildings are critical for providing a safe school and minimizing damage. For existing school buildings, this *Guide* provides information to school administrators and state officials to help identify potential safety problems and prioritize work to correct them. It also shares strategies that have worked to build support and funding for these potentially expensive projects. For new school buildings, the *Guide* shares best practices to ensure new buildings are as safe as possible and as likely as possible to reopen quickly after a hazard event.

- **Planning the Response (Chapter 4).** Every school should have an EOP that is inclusive of all individuals on the school campus and describes the actions that need to be taken before, during, and after an emergency occurs, who is responsible for each action, and contingencies for all of the different scenarios that could occur. This chapter provides an overview of EOPs and points to the natural-hazard specific aspects of developing such a plan.

- **Planning the Recovery (Chapter 5).** After the emergency timeframe is over, it can be a long road for schools to find their way to a new, post-disaster normal. This chapter presents detailed information on what schools should plan for and anticipate in advance to make recovery as positive and quick as possible after a destructive natural hazard event.

- **Engaging the Whole Community (Chapter 6).** In order to most effectively engage in comprehensive school safety planning, it is critical to engage all members of the community, including schoolchildren, parents, teachers, school administrators, school board members, emergency managers, and many other local leaders who can influence school safety decisions and activities.

Importance of Accounting for EVERYONE

"My oldest son is in a wheelchair, and he was in middle school in the valley when we had the Northridge (CA) earthquake. You know, there were a lot of aftershocks after that quake, but they never evacuated him with the rest of the students. He was on the third floor and they just left him there! They said they didn't have a plan for him." —parent (Barnes, 2013)

Planning with People in Mind

In Joplin, they made the multipurpose room for children with autism a tornado safe room (Freeman Health System, 2015). That way, the children with autism do not have to move if the event sirens sound as they are already in a safe space.

Learning from Hurricane Katrina

In Louisiana, many Individual Education Plans (IEPs) in paper form were destroyed during Hurricane Katrina. The state has since implemented a web-based system to store IEPs on state-operated servers. Access to all IEPs is now readily available to districts receiving special education students who transfer within Louisiana (National Forum on Education Statistics, 2010).

Taking Action Before an Event Strikes

Many schools in Oregon are of unreinforced masonry construction, which is particularly vulnerable to collapse in earthquake shaking. In response to this, concerned parents mobilized and as a group, advocated for earthquake retrofits in vulnerable schools. Over the years, their efforts have led to $175 million in state funding for school earthquake retrofit projects.

1.3.2 Goals of a Comprehensive Approach

As schools work to manage their natural hazard risk, it is important to have the work guided by a clear understanding of the ultimate goals of this effort. Clear goals both remind communities of *why* this work is so important and thus help to prioritize and focus efforts. Because there are many goals schools could have as they work to manage their risk from natural hazards, it is important that each school community determine the goals that are most meaningful and viable to them.

The following are some examples of typical goals for school hazard safety:

- **Safety.** Minimizing casualties among students, teachers, and staff is of paramount importance. This is especially important for schools because they work with children, who are dependent on adults to provide for their safety.

- **Emotional Well-Being.** Natural disasters are traumatic events for children and adults alike. Trauma is reduced when damage is minimized, emergency procedures are clear and practiced, and school routines resume quickly. Moreover, schools are important sites for identifying traumatized children and in providing mental health services or targeted interventions.

- **Educational Continuity.** Well-constructed and prepared schools can prevent or minimize occupancy interruption. When buildings are damaged, school contingency plans have to provide for education to temporarily resume elsewhere, which may be difficult in a large-scale disaster.

- **Savings and Benefits.** Money spent before a natural hazard to mitigate risk is a good investment that can save lives and help reduce the cost of repairing or replacing damaged buildings and building contents after an event. Studies funded by FEMA have shown a savings of $4 for every $1 spent on mitigation (Multihazard Mitigation Council, 2005).

- **Providing Emergency Shelter.** As some of the largest public buildings in many communities, school buildings often serve as evacuation sites, and/or post-event recovery centers. If used as an evacuation shelter, schools need to be properly designed and constructed as discussed in Chapter 3. Their loss can have repercussions to the community far beyond just education.

- **Speeding Community Recovery.** When students cannot attend school, parents may have a harder time going to work and contributing to helping the community get back to normal (see Chapter 5). School recovery is an important part of family and community recovery.

In addition to the goals listed above, it is important for school and community leaders to have conversations about why school safety is important, and then to set clear, tangible, and achievable goals for both the short- and the long-term. What are the requirements for school safety that need to be met? What are the goals for school safety and do those go beyond the requirements? What needs to be done to achieve those goals?

1.4 Requirements and Voluntary Measures for Schools

Building construction and emergency management requirements for schools vary by state and community. Some of the actions discussed in this *Guide* are required by law in some states and communities. Others are voluntary, advisable steps.

It is important for school leaders to familiarize themselves with specific requirements for their state and local community, in order to ensure proper

1-10   1: An Introduction to School Natural Hazard Safety   FEMA P-1000 and thorough compliance. Examples of some relatively common requirements include the following:

- School building construction generally falls under the jurisdiction of the state-adopted building code or is governed by statewide codes that are specific for school buildings. Most states follow the *International Building Code*, although this varies from state to state with some of the most hazardous regions adopting more stringent and state-specific building code construction and inspection requirements. Because building codes only provide the minimum necessary requirements to provide life safety and are not intended to control damage, "building above the code" to offer maximal safety and protection for occupants should be considered.

- Some states have special agencies, programs, or standards focused on safety in natural hazards for the design and construction of new school buildings. One example is the State of California's Field Act, which requires specific design review for new construction and special inspections for public schools (see the *Earthquake Supplement*).

- All states have requirements to provide adequate access and safety for persons with disabilities. These should be integrated into all aspects of a comprehensive school EOP.

- Some communities have laws requiring risk evaluation or retrofit for particular types of older school buildings. For example, several states in seismically-prone regions have laws that focus on unreinforced masonry school buildings—a particularly dangerous type of construction in earthquakes.

- Some states have laws requiring tornado shelters in new schools and the *2015 International Building Code* (ICC, 2014b) now requires the construction of tornado shelters in new schools with over 50 occupants in areas where the design wind speed for tornado shelters is 250 mph.

- Many states require local education agencies to comply with or provide assurances that schools create an EOP and conduct regular emergency drills.

- Some states require school emergency preparedness plans to be part of State Hazard Mitigation Plans.

Schools embarking on a safety strategy for exposure to natural hazards will need to understand which regulations and mandates apply to them. School leaders should refer to their State Department of Education, as it is most likely the primary source for identifying these requirements. It is important

> Life Safety is an engineering term used to describe a level of design. The main goal behind life safety is to prevent fatalities and serious injuries in a building due to failure or collapse of structural elements, such as columns and beams.

that schools work in close collaboration with their community partners who have roles and responsibilities in school preparedness, including law enforcement, fire officials, emergency managers, and mental health practitioners. School emergency management is a shared responsibility and community partners have expertise and resources to support schools.

Most states require basic emergency drills in schools, while others mandate stricter and more involved rigorous emergency planning. For state-specific emergency management resources, mandates, and partners, see the Readiness and Emergency Management for School (REMS) Technical Assistance website at http://rems.ed.gov/StateResources.aspx. An example of the type of information the website provides is illustrated in Figure 1-1.

STATE EMERGENCY MANAGEMENT RESOURCES

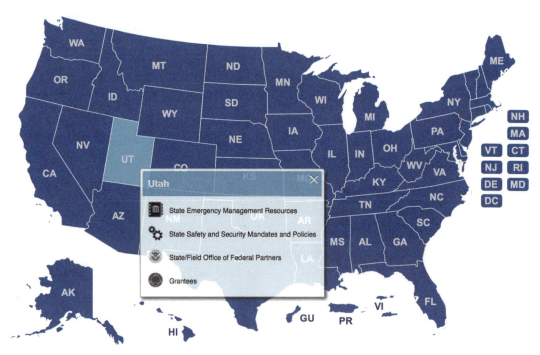

Figure 1-1   Types of state emergency management resources shown on the interactive map on the REMS website.

For a summary for legislative actions by state, see: csgjusticecenter.org/wp-content/uploads/2014/03/NCSL-School-Safety-Plans-Brief.pdf (Council of State Governments Justice Center, 2014).

Save the Children: 18 States Fall Short on Protecting Children in Disasters

Since 2008, Save the Children has issued annual reports on protecting children in disasters in the United States. As part of each report, the organization assesses the state of preparedness among United States schools and childcare centers. The 2015 disaster report card assesses whether states have met four criteria: (1) a plan for evacuating children in childcare; (2) a childcare plan for reuniting families after disaster; (3) a plan for children with disabilities and those with access and functional needs in childcare; and (4) a multi-hazard plan for K-12 schools. As of 2015, 32 states had met all 4 criteria, while 18 had not. Is your state prepared? Find out more at Save the Children: www.savethechildren.org

Figure 1-2   2015 disaster report card (Save the Children, 2015).

1.5 How to Use this *Guide*

This *Guide* is divided into three main sections, as shown in Table 1-1. This *Guide* only focuses on natural hazards. It does not cover other types of important hazards and threats for which schools should prepare, such as technological or biological hazards. It is not intended that users read all of the hazard-specific supplements—only those that cover relevant hazards should be read. Chapter 2 helps readers identify which natural hazards are relevant to their location.

Table 1-1   Organization of this *Guide*

Section	Content	Description
Comprehensive Approach for School Natural Hazard Safety	Ch 1: An Introduction to School Natural Hazard Safety Ch 2: Identifying Relevant Hazards Ch 3: Making School Buildings Safer Ch 4: Planning the Response Ch 5: Planning the Recovery Ch 6: Engaging the Whole Community Ch 7: Moving Forward	These chapters are recommended for all readers. They provide an overview of the key components of a comprehensive approach for school natural hazard safety.
Hazard-Specific Supplements	E: Earthquakes F: Floods H: Hurricanes TO: Tornadoes TS: Tsunamis W: High Winds X: Other Hazards: Snow Storms, Volcanic Eruptions, and Wildfires	These supplements provide detailed information and guidance focused on particular natural hazards. Readers should refer to the hazards that affect their schools.
Appendices	Earthquake Appendix Flood Maps Appendix Resources Appendix	The appendices provide more detailed information on a variety of topics, and are referenced in other sections of this *Guide*. The *Resources Appendix* points the reader to many sources of useful information that expand on the topics covered by this *Guide*.

Chapter 2

Identifying Relevant Natural Hazards

The key purpose of this chapter is to help readers identify which types of hazards could potentially impact their schools and determine which hazard-specific supplements they should be reading for more detailed guidance on addressing their risk. In particular, this chapter provides the following:

- An overview of the type of hazards that are covered in this *Guide*;

- A description of the characteristics of each of the hazards that are covered in this *Guide*, including a brief description of where they typically occur, their frequency of occurrence and intensity, warning time, duration, and follow-on hazards; and

- A summary checklist to help readers identify which hazard-specific supplement(s) apply to them. Hazard-specific supplements provide more detailed guidance on each particular hazard.

This information will help readers effectively work with the appropriate design professionals to identify, assess, and mitigate relevant risks to their school. It will also provide information on hazard characteristics (e.g., likely intensity and duration) that should be considered when developing school emergency operations plans and a risk reduction strategy.

2.1 An Overview of Natural Hazards

Some natural hazard events occur periodically during a person's lifetime. A community might remember recent floods or tornadoes. Other events, like earthquakes, may be infrequent by occurring perhaps once in several generations. Some communities may even experience multiple hazards. For example, they may be exposed to one type of natural hazard regularly, like flooding, and also have a major risk from infrequent events, like earthquakes or tsunamis. Regardless of their frequency, it is important for schools to be prepared for and ready to respond to all significant natural hazards that could reasonably be expected to occur.

Natural hazard events tend to occur repeatedly in the same geographical locations because they are related to physical or geological characteristics of an area or weather patterns. This is in contrast to technological hazards,

FEMA P-1000   2: Identifying Relevant Natural Hazards   2-1 biological hazards, and human-caused threats, which often refer to hazards borne from human action, whether accidental or intentional. Table 2-1 distinguishes the different hazard types, although it is worth remembering that hazards may be cascading, meaning that one hazard generates or is related to another (for instances, flooding and landslides may occur after a wildfire or oil spills may be generated after hurricanes). In addition, as scholars have long argued, a natural hazard only becomes a disaster when widespread community disruption occurs and many people are simultaneously affected. As shown in Table 2-1, this *Guide* focuses on six major natural hazards that are potentially devastating to schools: earthquakes, floods, hurricanes, tornadoes, tsunamis, and high winds. Other natural hazards, including snow storms, volcanic eruptions, and wildfires, are also briefly discussed.

Table 2-1   Natural Hazards and Other Types of Hazards and Threats (hazard categories adapted from U.S. Department of Education, 2013)

Type of Hazard	Specific Hazard	Extent Covered in this *Guide*
Natural Hazards	Earthquakes Floods Hurricanes Tornadoes Tsunamis High Winds	Covered in depth
	Snow Storms Volcanic eruptions Wildfires	Covered briefly
Technological Hazards	Includes: explosions or accidental releases of toxins from industrial plants; accidental releases of hazardous materials from within school (e.g., gas leaks, laboratory spills); hazardous materials releases from highways or railroads; radiological releases from nuclear power plants; dam failure; power failure; water failure	Not covered[1]
Biological Hazards	Includes: infectious diseases; contaminated food outbreaks; toxic materials present in school laboratories	
Adversarial, Incidental, and Human-caused Threats	Includes: fires; active shooters; criminal threats or actions; bullying, gang violence, or school violence; bomb threats; cyber-attacks; suicide	

[1] Covered in *School EOP Guide* (U.S. Department of Education, 2013).

It is important for schools to prepare for all hazards and threats that pose a potential risk, but this *Guide* focuses solely on natural hazards. Some resources listed in the *Resources Appendix* provide information and guidance that is also applicable to other types of threats and hazards that are not covered in this *Guide*.

2.2 Natural Hazards: Characteristics and Where They Occur

Each natural hazard has different characteristics that affect the types and amounts of damage it can cause, as well as the specific steps necessary to prepare. This section provides a brief description of each type of natural hazard covered in this *Guide*, and describes where they typically occur, their frequency of occurrence and intensity, warning time, duration, and follow-on hazards. These factors will play an important role in determining which hazards require immediate attention or more long-term planning and if mitigation efforts should address multiple hazards.

Characteristics of the natural hazards that can affect a given school should be taken into consideration when assessing building risk and mitigation of existing buildings, implementing best practices for the design of new buildings, and developing emergency operations plans that consider both response and long-term recovery.

Each hazard can have an effect over a length of time or have additional consequences following the event that should be considered when assessing the effects. Different parts of the United States experience different hazards more frequently than others.

This section is intended to be used by the reader to help determine which natural hazards may occur in their regions and how severe they may be. After identifying which hazards are relevant to their school, the reader should refer to each of the hazard-specific supplements in this *Guide*, which provide more detailed information and guidance that are particular to each hazard.

2.2.1 Earthquakes

Description. An earthquake is caused when two segments of the earth's crust suddenly slip along fault lines. This release of energy causes the ground to roll or shake and can cause damage or collapse of school buildings and property.

Where Earthquakes Occur. Studies indicate that there are 39 states within the United States with a significant earthquake hazard. Regions with the highest earthquake hazard include the western states, as well as some regions in the South and Midwest; areas with moderately high earthquake hazard include the Northeast. Figure 2-1 shows a map with relative seismic hazard

> Nearly half of the U.S. population is exposed to potentially damaging earthquakes, with about 28 million people living in areas with a high potential of experiencing damaging earthquake shaking (Jaiswal et al., 2015).

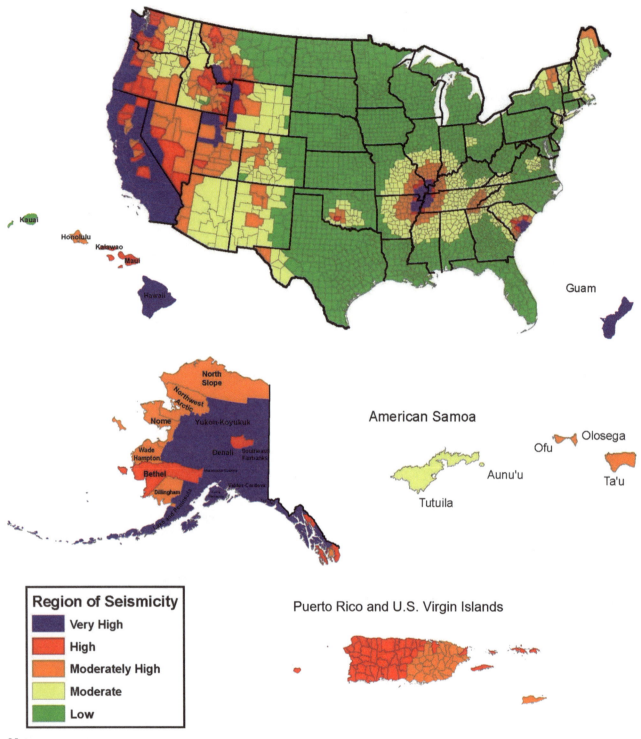

Notes:

(1) Based on NEHRP soil type B

(2) The seismicity at any site is calculated based on the highest seismicity at any point in a county. More accurate information on any site can be obtained from the USGS website: earthquake.usgs.gov/hazards.

(3) Islands not shown on American Samoa map (and their Region of Seismicity) are: (a) Rose Atoll (Low) and (b) Swains Island (Low).

Figure 2-1   Relative seismic hazard map showing earthquake intensity (FEMA, 2015b).

by state and territory of the United States. Schools that are in moderate, moderately high, high, and very high regions of seismicity should include earthquake risk in their school hazard safety plan. More information about earthquake hazard zones is available in the *Earthquake Supplement* and *Earthquake Appendix*, including a link to a website where schools can enter their address to get more detailed information on their particular earthquake hazard.

Frequency of Occurrence and Intensity of Earthquakes. Earthquakes can happen at any time in earthquake-prone regions. The interval between occurrences can vary widely and can only be estimated.

Earthquake size is categorized in various ways, the most commonly known measures are magnitude and intensity. The intensity of shaking at a given location depends on many factors, including earthquake magnitude, depth, and distance from the fault rupture, as well as soil characteristics at the site. Intensity is the best measure of the earthquake effects on school buildings. For example, a building on firm soil that experiences a distant, deep, and high magnitude earthquake will have less damage than a building on weak soil that experiences a nearby, shallow, and lower magnitude earthquake. Either way, earthquakes can impart tremendous loads on all elements of a building and generally result in some level of damage. Earthquake forces are so large that building codes are typically designed to allow a certain level of damage only because it is not economically feasible to design a building to not have any damage whatsoever.

> To learn more about the latest earthquake hazard maps and how to use them, visit: https://earthquake.usgs.gov/hazards/learn/.

Knowing how frequently earthquakes occur in a particular area and how strong they can potentially be can help determine the urgency for preparing for this type of hazard. Ultimately, in earthquake-prone regions, it is not a matter of if an earthquake will occur, it is a matter of when.

Available Warning Time for Earthquakes. At the time of writing, no operational warning system exists in the United States for earthquakes, but California and the United States Geological Survey (USGS) are in the process of developing such a system. When operational, this system could provide seconds to tens of seconds of warning before strong earthquake shaking reaches a given location. This short time frame would allow only the most basic life-saving steps, such as "Drop, Cover, and Hold On," to take place.

> Great ShakeOut Earthquake Drills are an annual opportunity for schools, along with their communities, to practice what to do during earthquake shaking. More information on the program can be found at www.shakeout.org

Duration of Earthquakes. Earthquake shaking lasts from seconds to minutes, depending on the type of earthquake, magnitude, and site location.

Duration plays an important role because longer duration of shaking can degrade the capacity of school buildings and lead to eventual collapse.

Follow-on Hazards from Earthquakes. Earthquake shaking can cause ground failure, landslides, fires, tsunamis, spills of hazardous materials, and failure of utilities and infrastructure. Earthquakes can also be followed by aftershocks, which can continue for weeks, months, or even years, and require additional consideration for safety during recovery.

2.2.2 Floods

Description. Flooding is a condition where water moves beyond normal channels and shorelines, and temporarily overflows and inundates normally dry areas. Floods can result from runoff from excess rainfall or snowmelt; ice or debris blockage of streams and drainage; high tides, waves, and storm surges; tsunamis; or failure of levees, flood protection structures, or dams. Floods can cause damage to school buildings and may require extensive repairs or even demolition.

Where Floods Occur. Flood prone regions include coastal areas and places subject to extreme rains or weather related events. Although all fifty states in the United States experience flooding hazards, the states with the most frequent and severe flood losses include Louisiana, Texas, New Jersey, New York, Florida, Mississippi, Pennsylvania, Alabama, North Carolina, South Carolina, Missouri, and California.

> Over 6,000 schools in the United States are located in a mapped Special Flood Hazard Area (SFHA) as defined by FEMA (Pew, 2017).

The most common flood hazard map is the Flood Insurance Rate Map (FIRM) produced by FEMA. FIRMs have been produced for over 21,000 communities in the United States. These maps show flood hazards from rivers, lakes, and the ocean and indicate Special Flood Hazard Areas (SFHA). In general, SFHA incicates an area with 1% annual chance of flooding, where the National Flood Insurance Program's (NFIP) floodplain management regulations and the mandatory purchase of flood insurance applies. More information about FIRMs is available in the *Flood Supplement* and *Flood Maps Appendix*.

> FIRMs can be viewed and downloaded from FEMA's Map Service Center (MSC) https://msc.fema.gov/portal. Maps can be searched by street address or by state and community. Flood hazard information can also be viewed using the National Flood Hazard Layer, accessible via the MSC.

Schools that meet any of the criteria listed below should include flood hazard in their school hazard safety strategy:

- schools located in SFHA (Zones A and V) on the FIRM;
- schools near the SFHA (Zones B, C, or X) on the FIRM;
- schools in areas behind levees;
- schools in hurricane storm surge inundation areas; or

- schools in areas that have experienced flooding in the past.

Past floods are also important given that flood hazard maps may not show local drainage issues or flooding from small watersheds. Local floodplain managers or local building or planning officials should be able to help with the determination of whether a school is in a flood zone.

Frequency of Occurrence and Intensity of Floods. Floods are the most common natural hazard in the United States, occurring in every state and territory. Flood intensity is often described in terms such as "100-year flood," which means a flood that has a 1% chance of happening in any given year. This does not mean that there will be 100 years between floods of this size.

Available Warning Time for Floods. Floods can occur with days of warning or, as in the case of flash floods, with little to no warning.

Duration of Floods. Flood inundation can last from hours to days to even months in some areas.

Follow-on Hazards from Floods. Floods can be accompanied by erosion, mudslides, debris flow, and high velocity waves. Floods can cause spills of hazardous materials, fires, and failure of utilities and flood protection structures.

2.2.3 Hurricanes

Description. A hurricane is a tropical weather system of spiraling winds converging with increasing speed toward the storm's center (the eye of the hurricane). Hurricanes form over warm ocean waters. Many hurricanes bring very high winds and heavy rainfall and/or coastal flooding. Hurricanes also occasionally spawn tornadoes. The terms "hurricane," "cyclone," and "typhoon" describe the same type of storm. The term used depends on the region of the world where the storm occurs.

Where Hurricanes Occur. Hurricane hazard areas in the United States include Atlantic and Gulf Coast areas, Hawaii, and U.S. territories in the Caribbean and South Pacific, which includes American Samoa, Guam, Northern Mariana Islands, Puerto Rico, and U.S. Virgin Islands. Regions along the Atlantic and Gulf coasts that are at risk of hurricanes are shown in Figure 2-2 (red and orange areas on the map). Schools located in these regions, or in Hawaii or United States territories in the Caribbean or South Pacific should include hurricane risk in their school hazard safety strategies. The *Hurricane Supplement* provides more information.

Figure 2-2   Map indicating hurricane hazard areas in the United States (high wind area adapted from ASCE, 2017b).

Frequency of Occurrence and Intensity of Hurricanes. In the Central Pacific, Atlantic, Caribbean, and Gulf of Mexico regions, hurricane season runs from June through November. During typical hurricane seasons, about six hurricanes can be expected. During more active years, up to 15 hurricanes may occur along the United States Gulf or Atlantic coasts (NOAA, 2015). In the West Pacific region, hurricanes can occur in every season.

The Saffir-Simpson Hurricane Wind Scale categorizes the intensity of hurricanes based on wind speed. The five-step scale ranges from Category 1 (the weakest) to Category 5 (the strongest). Typical hurricanes are about 300 miles wide, although they can vary considerably. The largest one on record had a diameter of 1,350 miles and the smallest was 60 miles.

Available Warning Time for Hurricanes. Schools typically have more than a day of warning time before a hurricane strikes.

Duration of Hurricanes. Extremely strong winds can last for several hours and moderately strong winds for a day or more. Flooding associated with hurricanes can persist for hours (storm surge) or up to days or weeks (rainfall induced flooding).

Follow-on Hazards from Hurricanes. Hurricanes can also spawn tornadoes and are often accompanied by storm surge flooding with waves and flooding caused by torrential rain. This can lead to spills of hazardous materials and failure of utilities and infrastructure.

2.2.4 Tornadoes

Description. A tornado is a violently rotating column of air extending from the base of a thunderstorm to the ground. Although small in terms of impact area, tornadoes can generate wind speeds that are far greater than what schools are typically designed to resist.

Where Tornadoes Occur. Tornadoes can occur throughout the United States. However, the most destructive, deadly, and strong tornadoes mostly affect central United States, and are rare in the West and Northeast. The shaded areas in Figure 2-3 indicate the portion of the continental United States that is considered to be in a tornado-prone region and should include tornado risk in their school hazard safety strategy. For locations along or near a boundary of different wind speeds on the Figure 2-3 map, the higher wind speed should be assumed. The *Tornado Supplement* provides more information.

Frequency of Occurrence and Intensity of Tornadoes. On average, over 1,000 tornadoes are recorded in the United States each year. Tornadoes can occur at any time during the year and at any time in the day; however, different regions are more likely to experience tornadoes at certain times of the year. Tornadoes in the "Tornado Alley"—a nickname given to a region in the southern plains of the central United States that consistently experiences tornadoes each year—are more likely to occur in late spring and sometimes in the early fall. Gulf states are more likely to experience tornadoes earlier in the spring, and in the northern plains and upper Midwest, summer is more likely. Fewer tornadoes are documented in winter months, although deadly winter outbreaks have occurred. In terms of time of day, tornadoes are most likely to occur mid-afternoon to evening, although they can occur at all hours of the day (NOAA, 2017).

Tornadoes are typically less than 1,000 feet wide; however, widths of approximately 2.5 miles have been reported. The National Weather Service

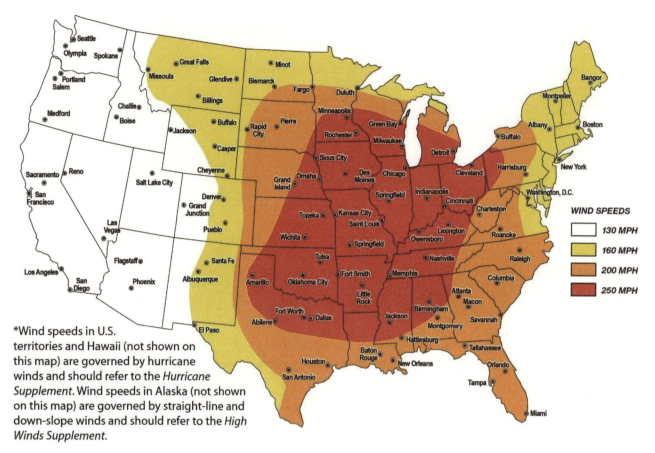

Figure 2-3   Map indicating tornado-prone regions in the United States. Schools that are located in the 160, 200, and 250 miles per hour wind speed zones are considered to be in a tornado-prone region (adapted from ICC, 2014a).

rates tornado severity using the six-level Enhanced Fujita (EF) scale, with EF0 (least severe) to EF5 (most severe), based on observed damage.

Available Warning Time for Tornadoes. Tornadoes occur with a few or several minutes of warning.

Duration of Tornadoes. Tornado winds generally last several seconds to a minute at a given location. In total, tornadoes can last from several seconds to more than an hour, with most lasting less than 10 minutes (NOAA, 2017).

Follow-on Hazards from Tornadoes. Tornadoes can rupture natural gas lines resulting in fires. Tornadoes can also cause spills of hazardous materials and failure of utilities and infrastructure.

2.2.5   Tsunamis

Description. A tsunami is a series of waves that are caused by a rapid disturbance within a body of water. Waves travel outward in all directions from an initial tsunami generating source, usually from an undersea

> **Watch vs. Warning**
>
> A tornado watch is issued when conditions are favorable for tornado development.
>
> A tornado warning is issued when a tornado has been sighted or indicated by weather radar.

earthquake, much like the ripples caused by throwing a rock into a pond. Because tsunamis are actually a series of pressure waves, they can travel across oceans at great speed and increase to significant heights as they come onshore. They can cause low-lying areas adjacent to coastlines to experience severe inland inundation of water and debris, causing significant damage to school buildings and property.

Where Tsunamis Occur. Although all coastal areas in the United States can experience a tsunami, the regions with the highest tsunami hazard level include Hawaii, Alaska, and the West Coast states (especially the Pacific Northwest states). American Samoa, Guam, Northern Mariana Islands, Puerto Rico, and the U.S. Virgin Islands are considered to have a high hazard level, as well (see Table 2-2).

Through the National Tsunami Hazard Mitigation Program (NTHMP) that is administered by the National Oceanic and Atmospheric Administration (NOAA), many coastal states and territories have developed tsunami inundation maps and evacuation routes with signage that depict where tsunami inundation may occur and how people can evacuate to high ground. Schools located in regions with high or very high hazard levels per Table 2-2 should refer to the *Tsunami Supplement* for more information.

Table 2-2   Coastal Areas in the United States Ranked by Tsunami Hazard (adapted from NTHMP, 2015)

Location	Tsunami Hazard Level
Alaska	High to Very High
Hawaii	High to Very High
U.S. West Coast	High to Very High
American Samoa	High
Guam & N. Mariana Islands	High
Puerto Rico & U.S. Virgin Islands	High
U.S. Atlantic Coast	Very Low to Low
Alaska Arctic Coast	Very Low
U.S. Gulf Coast	Very Low

Note: Hazard levels are qualitative and based largely on the historical record from the early 19th century through 2014, geological evidence, and location relative to known tsunami sources, all of which provide clues to what might happen in the future.

Frequency of Occurrence and Intensity of Tsunamis. A damaging tsunami occurs approximately twice a year worldwide. The frequency of tsunamis in a particular location mostly depends on how often offshore faults produce earthquake-generating tsunamis. For example, tsunamis are more likely to occur along Pacific Ocean coastlines due to the number of active undersea earthquake faults around the Pacific Rim. Tsunamis can also be generated by other sources, such as underwater landslides, but these events are rarer and more difficult to forecast.

Tsunamis are measured by the wave height above normal sea level (amplitude), the on-land depth of flooding (inundation) and the distance they penetrate inland (runup).

Available Warning Time for Tsunamis. For a locally-generated tsunami, earthquake shaking is "nature's warning sign" and provides several minutes of warning prior to arrival of a tsunami. For a tsunami generated by a distant source, there can be hours of warning time provided by a Tsunami Warning Center under NOAA's Tsunami Program. Tsunami warnings, advisories, and watches can be received via the NOAA Weather Radio.

Duration of Tsunamis. Tsunami waves come as a series of waves, and can last for up to 24 hours although the most damaging waves usually occur within a few hours' time.

Follow-on Hazards from Tsunamis. Tsunamis bring severe flooding with significant debris and can cause fires, spills of hazardous materials, and failure of utilities and infrastructure.

2.2.6 High Winds

Description. Most wind damage is caused by tornadoes and hurricanes. However, damage is occasionally caused by other high winds, notably straight-line and down-slope winds. Straight-line winds generally blow in a single direction and are common throughout the United States. Down-slope winds blow down the slope of mountains.

Where High Winds Occur. Straight-line winds with sufficient speed to cause building damage can occur anywhere in the United States and its territories. Down-slope winds with sufficient speed to cause building damage can occur in mountainous areas. Because these high winds can occur anywhere, all users who are not already reading the *Hurricane Supplement* should read the *High Winds Supplement*.

Frequency of Occurrence and Intensity of High Winds. Damaging straight-line and down-slope winds can occur at any time.

Available Warning Time for High Winds. Weather events generating straight-line and down slope winds can generally be predicted days or hours in advance.

Duration of High Winds. Winds associated with intense low pressure can last up to a day at any given location.

Follow-on Hazards from High Winds. Other hazards associated with high winds are wildfires and snow drifts, which are snow mounds created by high winds.

2.2.7 Other Hazards

This *Guide* also includes brief information about the following other natural hazards: snow storms, volcanic eruptions, and wildfires. Additionally, there may be natural hazards not specifically mentioned in this *Guide*, like droughts or hailstorms, which should be considered when developing a comprehensive strategy to address school natural hazard safety. If these hazards are known to be relevant to a school, they should be incorporated into school hazard safety strategies. More information is provided in the *Other Hazards Supplement*.

2.3 Summary Checklist – Which Hazards are Relevant to Your School?

1. Take the following steps to determine whether your school has a reasonable chance of experiencing each of the hazards below:

   □ **Earthquakes:** Is your school in a moderate Region of Seismicity or higher per Figure 2-1? If so, read the *Earthquake Supplement*.

   □ **Floods:** Is your school located behind a levee, in a storm surge inundation area, or in Flood Zone A, V, B, C, or X? Does your school have a history of flooding? If any of these are true, read the *Flood Supplement* and the *Flood Maps Appendix*.

   □ **Hurricanes:** Is your school in the shaded region in Figure 2-2 or in Hawaii or a U.S. territory in the Caribbean or South Pacific? If so, read the *Hurricane Supplement*.

   □ **Tornadoes:** Is your school within the tornado-prone region as defined in Figure 2-3? If so, read the *Tornado Supplement*.

   □ **Tsunamis:** Is your school within a high or very high tsunami hazard level per Table 2-2? If so, read the *Tsunami Supplement*.

   □ **High Winds:** All areas in the United States are susceptible to high winds, notably straight-line and down-slope winds. If you are not already reading the *Hurricane Supplement*, you should read the *High Winds Supplement*.

   □ **Other Hazards:** If you think your school is located in an area that is prone to snow storms, volcanic eruptions, or wildfires, read the *Other Hazards Supplement*.

2. Incorporate risk management steps for the relevant hazards in your school's hazard safety strategy.

FEMA P-1000 **2: Identifying Relevant Natural Hazards** **2-13**

Chapter 3

Making School Buildings Safer

School buildings vary tremendously in characteristics, size, age, condition, and construction materials. For instance, schools can range from one-room schoolhouses to large campuses with multi-story complexes. School construction materials can vary from wood, concrete, steel, masonry or a combination of these. Some schools are located in dense urban environments, while others are in rural or suburban settings. In addition to their setting, the location of school buildings also dictates to which natural hazards they might be exposed, as was covered in Chapter 2.

> Well-designed, constructed, and maintained school buildings are critical for providing a safe and reliable learning and work environment.

Given all these varying factors, each individual school will have a unique situation and will require a corresponding mitigation plan that is particular to the specific context and situation. Although the specifics of mitigation plans are unique, the approach and strategies for making school buildings safer are rooted in the same basic process.

The purpose of this chapter is to provide guidance on improving the structural safety and resiliency of school buildings. In particular, this chapter provides:

- An overview of the level of safety from natural hazards that is expected and provided by buildings codes for school buildings;

- Guidance on determining vulnerabilities of existing school buildings, evaluating mitigation options, and developing an implementation plan;

- Guidance on important considerations for new school building construction to improve natural hazard resilience;

- Suggestions for developing a funding plan for mitigation work or new school facilities that incorporate natural hazard resilience; and

- A description of important quality assurance measures that are necessary to provide a safe school facility long-term.

3.1 School Building Safety from Natural Hazards

Buildings can typically withstand common weather events, such as rain, snow, and wind. However, infrequent, but strong natural hazard events bring

FEMA-P-1000   3: Making School Buildings Safer   3-1

forces that put enormous loads and stress on buildings, which can lead to damage or even collapse.

Different types of natural hazards affect buildings in different ways. For instance, earthquake shaking could cause damage to the entire structural framework rendering it unusable whereas a hurricane or tornado may only affect a section of the roof or wall while leaving other areas marginally affected and potentially usable.

While avoiding damage to school buildings from any natural hazard is desired, it is particularly important for schools that are prone to hazards that provide little to no warning. Because warning times can vary, some students and staff could still be inside school buildings during an event, whereas for other hazards the school is likely to have been evacuated and the building would be empty during the time of the event. Table 3-1 provides a list of the likely warning times by hazard and the corresponding expected occupancy if the event were to strike during school hours.

Table 3-1   Likely Warning Times by Hazard and Resulting School Occupancy Expectation During Hazard Events

Hazard	Expected Warning Time	Expected Occupancy in a Hazard Event during School Hours
Earthquake	No warning or seconds of warning	Occupied, given the lack of warning time
Flood	Usually hours to days of warning; sometimes no warning, especially for flash floods	Evacuation in advance is likely in most cases Occupied in rare cases, such as flash floods
Hurricane	Days of warning	Evacuation in advance is likely in most cases Occupied if the school building is a designated hurricane evacuation shelter
Tornado	Minutes of warning	Occupied, given the lack of warning time In some cases, the school building will have a designated tornado safe room
Tsunami	Minutes of warning for local tsunami Hours of warning for distant tsunami	Evacuation in advance is likely in most cases Occupied if the school building is a designated tsunami vertical evacuation building
High Winds	Varies, minutes to hours of warning	Typically occupied, given the lack of warning time in most cases

For schools threatened by hazards with little or no warning, building assessments and evaluations should be prioritized. For school buildings exposed to hazards with longer warning times, effective evacuation procedures and continuous training is particularly important. In all cases, it is important to have school buildings that are hazard resistant to minimize damage and interruptions, and to ensure educational continuity.

3.1.1 Level of Safety Provided by Building Codes

Many people assume that the government requires all school buildings to be safe and minimally damaged during a natural hazard. This is not necessarily the case. Most local governments do require that new schools be designed and constructed to meet local building codes, which are generally based on current state and national model codes. In some states, school construction may be governed by a statewide school construction code, which may differ from the building code used for other types of buildings. In many cases, the objective of the adopted building codes is to provide life safety at a minimum for some hazard events, but not necessarily prevent structural damage to the school building. In fact, if a school is struck by a strong, violent tornado or earthquake, significant damage is expected even if the school building was constructed using modern day building codes. Put simply, just because a building is "built to code" does not necessarily mean that it will be fully functional or usable after a hazard event. Additionally, after a school building is constructed, structural changes over the years to maintain or enhance resistance to natural hazards are typically not required.

> Life Safety is an engineering term used to describe a level of design. The main goal behind life safety is to prevent fatalities and serious injuries in a building due to failure or collapse of structural elements, such as columns and beams.

Building codes improve over time with respect to natural disaster resistance as experts learn from hazard events and building science research. In fact, substantive building code changes, standards, and test methods have been made since the 1990s to reflect the notion that buildings should perform better in hazard events. This means that older school buildings are designed and constructed to older building codes that do not reflect modern knowledge about safe building design and construction. Consequently, many older school buildings are significantly less resistant to natural hazards than schools constructed to current building codes. Some older school buildings might have safety risks that are unacceptable to the community.

> How old are your school buildings?
>
> Over 40% of school buildings in the United States are over 15 years old (U.S. Department of Education, 2012), meaning that they were not designed and constructed to the latest building codes and standards, which require that schools be designed to withstand stronger loads. The average public school building is over 40 years old (NCES, 2014).

In addition, many community leaders now argue that aiming for only life safety in school buildings is not adequate, as this goal does not prevent damage that could render the building unusable after some hazard events. A better objective, especially for new school buildings, is immediate occupancy. This means designing, constructing, and maintaining school buildings so they do not suffer significant damage and are more likely to be

---

FEMA-P-1000   3: Making School Buildings Safer   3-3

usable again shortly after an event. This objective aims to minimize disruption and improve community resilience given that the resumption of school is closely tied to community recovery. Because schools are often planned to be used as community emergency shelters, they should be designed and built to be functional following a natural hazard event.

Cascadia Earthquake and Tsunami Risk Looms Large

The Cascadia subduction zone, widely referred to as the Cascadia fault, runs from northern Vancouver Island in Canada to northern California. The last known great Cascadia earthquake took place in 1700. Although hundreds of years have passed with no major activity along the fault line, many cities are at risk including Vancouver and Victoria, British Columbia; Seattle, Washington; and Portland, Oregon. Victoria, Vancouver, and Seattle are coastal cities that also face tsunami risk.

Across Washington State, about 386,000 students—or one in every three enrolled—live in earthquake-prone areas and attend schools built before seismic construction standards were adopted statewide. In addition, about 31,000 students in Washington attend schools that are in tsunami inundation zones (Doughton and Gilbert, 2016).

No one knows if the next "big one" along the Cascadia fault will occur in our lifetimes. But just because these events are infrequent, it does not mean that they can be ignored. The lives of children and those who teach them on a daily basis depend on action.

3.2 Existing School Buildings

School leaders interested in addressing their school's potential natural hazard risk should follow the following steps: (1) engage a team of qualified engineers and architects to determine school building vulnerabilities; (2) identify and evaluate mitigation options and corresponding costs; and (3) develop a plan to fund and implement mitigation actions. In some cases, a long-term program to fund and implement building improvements may be needed.

3.2.1 Determining Building Vulnerability

Each natural hazard affects school buildings in different ways. Earthquakes shake the entire building intensely, hurricane and tornado winds produce pressures on the exterior wall and roof systems, and floods and tsunamis generate tremendous water pressures on exterior portions of the building. Although these considerations require different analytical approaches, the overall process for identifying and mitigating significant building vulnerabilities is essentially the same and will require the advice of design professionals, as well as the expertise of school facility and financial managers.

3-4   3: Making School Buildings Safer   FEMA-P-1000

Some school leaders are reluctant to examine the natural hazard vulnerability of their school buildings for a variety of reasons. They worry that any needed upgrades or retrofits will be too expensive. They fear the reaction of parents whose children attend the school if they learn the building is potentially unsafe in a natural disaster. Further, the authority and responsibility to evaluate and address structural vulnerabilities in existing school buildings is often unclear and is typically not explicitly part of anyone's job description.

Nevertheless, none of these reasons for reluctance change the fundamental importance of school buildings having adequate resistance to natural hazards. In many communities, emergency management champions have emerged who are willing to face these challenging issues for the greater good of the community and its children.

National standards exist that define consistent, technically rigorous approaches to assess building vulnerability for most natural hazards. Details about how to conduct building vulnerability assessments for each natural hazard are provided in the hazard-specific supplements in this *Guide*. For each hazard, the specific process of identifying vulnerabilities and defining risk reduction options is different. In most instances, a team of qualified structural engineers and architects is needed to make these assessments. A quick, low-cost first step is to have the experts identify some preliminary characteristics, such as building age, type of construction, and hazard exposure at the school site. This preliminary screening information allows the team to quickly identify possible red flags so that school leaders can have a deeper conversation about potential next steps for mitigation.

> The following provides case studies of successful school earthquake screening programs in the United States: www.eeri.org/projects/schools/subcommittees/#eval.

Understanding a school building's vulnerabilities is particularly important if it is pre-designated as a shelter for people to use during a natural hazard event. For example, if a building is designated as a safe place to go during a hurricane or tsunami, it should be ensured that the building is designed and constructed to protect people inside the building during such an event. Some school buildings might also be designated to be used after an event to help with recovery. In those cases, it is particularly important to understand if the building is likely to be functional following a natural hazard event. If any of these designations are applicable to a school, it is critical for the team of qualified professionals conducting the building assessment to be aware of this. More information on understanding a school's role as an emergency or recovery shelter is provided in Section 5.1.5.

Designating Adequate Buildings as Shelters is Critical

On September 21, 1989, Hurricane Hugo, by then a Category 4 storm, made landfall on the South Carolina coast above Charleston. The Town of McClenllanville had a single approved shelter for its residents—the local high school. Unfortunately, the designation was based on erroneous elevation of the school building. When the storm reached its peak, people had to stand on desks, break through the ceiling, and place their children above to avoid drowning. This event highlighted the importance of ensuring that pre-designated shelters will remain safe to occupy during natural hazards.

3.2.2 Identifying and Evaluating Mitigation Options

Mitigation options can range significantly in their scope, cost, implementation time, level of disruption, required personnel, and effectiveness at reducing risk.

Mitigation options should also be evaluated in light of building code requirements for improvements and repairs. Significant improvements or alterations can trigger requirements for the work to comply with the current building code, which could include additional considerations, such as accessibility and energy improvements. Most work to strengthen buildings from natural hazards will be viewed favorably by building departments provided that the overall building is not made less strong. Additionally, repairs after an event can also trigger building code requirements based upon the amounts of damage to the building. Relevant requirements should be discussed and evaluated with the professional design team before committing to the work. Section 5.1.3 provides more information on these trigger requirements.

3.2.3 Developing an Implementation Plan

After school building vulnerabilities and mitigation options are identified, the challenging tasks of prioritizing and phasing actions, building community and political support, and raising needed funds begin. This can be a difficult and uncertain process. Seeking funds for facilities improvements, replacement, or new construction is a tough financial task most school leaders will eventually face. Even the smaller funds needed for planning activities and engineering assessments that precede capital projects can sometimes be unavailable or difficult to raise.

School leaders with responsibility for large numbers of buildings will likely need to prioritize mitigation activities. It often makes sense to prioritize projects based on a "worst first" concept. This strategy first addresses the most pressing vulnerabilities affecting the largest number of children,

Earthquake Mitigation Examples

A school with earthquake vulnerabilities might consider strengthening to improve its expected performance in an event in conjunction with other scheduled maintenance and upkeep tasks. This often can be accomplished over periods when the school is unoccupied (summers) and integrated incrementally over a number of years to help reduce disruption costs. FEMA 395, *Risk Management Series: Incremental Seismic Rehabilitation of School Buildings (K-12)* (FEMA, 2003), provides more detailed guidance on applying these strategies to reduce seismic risk in schools.

Teachers and students can conduct a hazard hunt for potential falling hazards, such as unsecured bookshelves and light fixtures, in a school with earthquake risk. These risks can often be easily improved and can be cost effective, such as adding of straps and anchors. FEMA's Earthquake School Hazard Hunt Game and Poster is a great resource: https://www.fema.gov/media-library/assets/documents/ 90409. For more details on earthquake-specific mitigation options, see the *Earthquake Supplement*.

Flood Mitigation Examples

A school with flood risk can elevate or dry floodproof the structure to protect the building and its contents from flood damage. This should include the utility equipment such as furnaces, boilers, and air conditioning so that all portions of the building survive flooding with minimal damage. A school building can be elevated using compacted soil in low flood velocity areas or columns or pilings in areas where flood velocities are higher. In addition, areas such as entrances and storage rooms located below flood elevations should be constructed of flood-resistant materials such that they can be easily cleaned and repaired after a flood event. For more details on flood-specific mitigation options, see the *Flood Supplement*.

Hurricane Mitigation Example

A school replacing a roof covering that has reached the end of its service life in a hurricane-prone region could incorporate best practices in the design and construction of the new roof. For more details on hurricane-specific mitigation options, see the *Hurricane Supplement*.

Tsunami Mitigation Example

A school with tsunami risk might determine that it is not feasible to reach safe ground from the school premises within the expected tsunami warning time. The school could construct a new wing of the school that can both resist the forces of tsunami inundation waves and is tall enough to be used for vertical evacuation. This means the building's highest floors are higher than expected tsunami inundation waters. For more details on tsunami-specific mitigation options, see the *Tsunami Supplement*.

Tornado Mitigation Example

When designing an addition to an existing school building, a school with tornado risk can incorporate a tornado safe room into the addition which would be large enough to accommodate all students, staff, visitors to a school, as well as nearby community members whose residences do not have suitable shelter. For more details on tornado-specific mitigation options, see the *Tornado Supplement*.

followed by less critical vulnerabilities. In some cases, mitigation is incorporated as part of planned maintenance or replacement of deteriorated building components. In other cases, communities pursue less expensive projects first, while they build support and raise funds for more expensive projects. Section 3.4 provides recommendations on developing a funding plan.

Keeping Students and the Community Safe

After nearly one-third of the town of Joplin, Missouri, was flattened when a massive EF5 tornado touched down, community leaders were dedicated to build back better. According to FEMA, the number of tornado safe rooms in Missouri has doubled since the Joplin tornado in May of 2011. The City of Joplin had 14 community tornado safe rooms as of May 2016, many of them located inside local schools. This will ensure that students, teachers, and members of the public have a safe place to go in the event of a tornado. (McTavish, 2016)

FEMA can help fund the construction of safe rooms through their Hazard Mitigation Grant Program (HMPG). Section 3.4 provides more information on HMGP and other financial resources. The Pre-Disaster Mitigation Grant Program can also provide support even if a community has not experienced a recent disaster.

Figure 3-1   Tornado safe room placard.

Existing School Buildings: Summary of Key Steps

- Determine existing building vulnerabilities by engaging a team of qualified engineers and architects to conduct school buildings assessments.

- Identify and evaluate mitigate options to address all significant vulnerabilities.

- Develop a plan to implement building mitigation actions over time, depending on the needed scope and cost, with participation of community stakeholders.

3.3 New School Buildings

The design and construction of new school buildings provides communities with the opportunity to "get it right" when it comes to disaster resilience. This requires careful decision-making, input at the earliest stages of the project, and oversight throughout the building process and ultimately results in a school building that can serve the community well for decades to come. In many cases, new school buildings can be made highly natural hazard-resistant with only minimal increases in design and construction costs compared to a typical new school building. For example, elevating the new building above the minimum flood elevation required by the building code may add minimal cost to the project, but significantly improve flood resilience and lower flood insurance premiums. Additionally, initial design and construction decisions can have a significant effect on operational and maintenance expenses over the lifetime of the school building.

> For new school buildings, adding more hazard resistance typically only increases construction costs a small amount. For example, studies have shown that providing adequate seismic design generally adds less than 2% to the overall cost of typical building construction (NIST, 2014).

In order to have new school buildings that are resistant to natural hazards, the following should be considered: (1) smart site selection; (2) relevant building codes and resilient design; and (3) schools as emergency shelters or recovery centers.

3.3.1 Smart Site Selection

Selecting sites for new construction that are less prone to natural hazards is highly advised. For communities with flood and tsunami risks, new school buildings should be located outside of mapped hazard zones. In areas with seismic risk, careful consideration of soils, landslide potential, and proximity to fault lines should be considered when selecting sites for new schools.

Chapter 2 provides an overview of where specific natural hazards typically occur, and more details are provided in the hazards-specific supplements.

3.3.2 Relevant Building Codes and Resilient Design

First, it is important to understand what level of damage is expected in new school buildings designed to current codes. New schools must be planned, sited, designed, and constructed in accordance with the state building codes that are based on national standards. In some states, school construction may be governed by a state school construction code, which may differ from the building code used for other types of buildings.

In general, building codes require that school buildings be designed to be somewhat stronger than typical buildings (e.g., residences or commercial spaces). However, as described in Section 3.1.1, the minimums prescribed in buildings codes only provide a certain level of protection, which some

---

FEMA-P-1000   3: Making School Buildings Safer   3-9

communities think is not adequate for schools. Because of this, some school leaders have chosen to design and construct new school buildings to go beyond the code. By doing so, they improve the resiliency of their schools and their communities. Often, significant improvements in building resilience can be achieved for relatively small additional costs. For instance, the addition of a modest amount of reinforcing steel at a nominal cost can significantly improve the performance of foundation or structural systems.

Investing in more resilient school buildings pays off—it can save lives and help reduce the cost of repairing or replacing damaged buildings. The costs are small when compared to the benefits.

3.3.3 Schools as Emergency Shelters or Recovery Centers

Communities also often desire to use school buildings as shelter spaces during and after events. This can include providing enough space in tornado safe rooms to accommodate community members, or making a school an official hurricane evacuation shelter. It can also include using school spaces for community recovery functions after an event. If local emergency officials and community leaders expect for a school building to serve as an emergency shelter or recovery center, it should be discussed and addressed in the conceptual design phase of a new school building. School buildings that are designated shelters or recovery centers have additional design requirements, such as designing to resist higher loads and protecting certain equipment, to ensure that they will remain functional during and after events.

Decisions related to whether or not the school building should serve as a shelter should be made within the context of the entire community, its needs, and corresponding costs. For example, school districts might want to make strategic decisions by selecting a few, centrally-located school buildings to design as shelters if they cannot afford this level of design for all schools.

Many of the emergency shelter and recovery center requirements vary significantly by the hazard being addressed. For example, the design requirements for a tsunami vertical evacuation shelter can be quite different from that of a tornado safe room. More specific considerations related to this topic are provided in the hazard-specific supplements in this *Guide*.

> **New School Buildings: Summary of Key Considerations**
>
> - Identify appropriate sites for new schools considering hazard-specific concerns.
>
> - Understand the relevant building codes and the corresponding level of safety that they provide, and decide if the new school design should go beyond the code requirements.
>
> - Determine if the new school will also serve as a designated emergency shelter or recovery center.

3.4   Developing a Funding Plan

The necessary political support and funding depend on the scope and cost of work to be done. As a start, State Hazard Mitigation officers are a great resource to obtain information about federally funded projects and their associated costs. Projects that entail major work on existing buildings or construction of new buildings can be very expensive. However, such expenditures can save lives and reduce the costs associated with a school that is heavily damaged or destroyed by an event. For expensive projects, schools can raise needed funds in a variety of ways, including the following:

- school bond measures or other ballot initiatives,

- special fundraising campaigns online or through community, and

- public-private partnerships between schools and companies or corporations that may be willing to donate expert time, materials for construction, or other goods.

As far as federal support, school districts should consider participating in hazard mitigation planning processes conducted by local jurisdictions or even developing a FEMA-approved Natural Hazard Mitigation Plan of their own. With such a plan in place, districts become eligible for two federal grant programs that can be used to support mitigation options, such as strengthening an existing school structure. These programs are:

> **FEMA Grant Opportunities for School Natural Hazard Safety Improvements:**
>
> **Before A Disaster:** Pre-Disaster Mitigation (**PDM**) Grants
>
> **After A Disaster:** Post-Disaster Hazard Mitigation Grants (**HMGP**)
>
> More information can be found here: https://www.fema.gov /hazard-mitigation-assistance.

- Pre-Disaster Mitigation (PDM) Grant Program; and

- Hazard Mitigation Grant Program (HMGP) for post-disaster.

Both of these FEMA funding programs are administered through state emergency management agencies, with grants disbursed to jurisdictions or public entities like school districts that have a Natural Hazard Mitigation Plan in place. FEMA planning grants may be available to defray the expense of preparing such a plan.

School districts often make plans to spread mitigation work over a multi-year timeframe. This type of work can also be incorporated into annual

---

FEMA-P-1000   **3: Making School Buildings Safer**   **3-11**

School District's Hazard Mitigation Plan Secures Investment in Tornado Safety

Beggs Public Schools, 35 miles south of Tulsa in northeastern Oklahoma, sought to augment safety for the district's 1,201 students and 166 teachers and staff in the event of severe tornadoes. The multi-million-dollar cost of a new facility designed to high safety standards was beyond the means of the small rural school district. Former Superintendent Cindy Swearingen (now retired) wondered if federal dollars could help.

The Superintendent started by asking Okmulgee County Commissioners to add a tornado safe room project to the county's existing Hazard Mitigation Plan in order to secure eligibility for federal grant funds. She learned that lacking planning grant dollars, the county was unable to update its plan to accommodate the school project.

Superintendent Swearingen then approached her school board and requested $20,000 to hire a consultant to prepare a Hazard Mitigation Plan specifically for the district. The board approved her request and hired a writer. As soon as the Oklahoma Department of Emergency Management and the Federal Emergency Management Agency approved the district's first Hazard Mitigation Plan, the district prepared and submitted its grant request to the state's Hazard Mitigation Grant Program (HMGP), which is ultimately funded by FEMA.

In the wake of the catastrophic EF5 Moore tornado near Oklahoma City in May 2013, a Presidential Disaster Declaration released a large amount of new federal funding to the State of Oklahoma's HMGP. Beggs School District's grant was not funded at that time, however. After three years and more disaster declarations replenished the HMGP, Beggs Schools successfully secured $3 million toward the $4 million expense of a new dual-purpose building containing school band facilities and a community tornado safe room.

When completed, the building will be a monolithic dome designed to withstand an EF5 tornado. Thanks to a federal grant, a new dome resembling the existing Beggs Event Center (Figure 3-2) will house Beggs School District's tornado safe room.

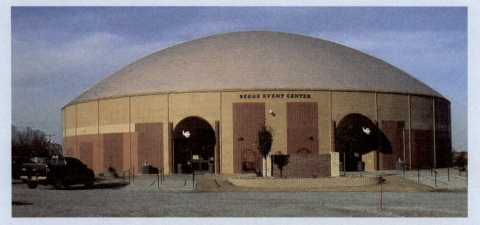

Figure 3-2 Thanks to a federal grant, a new monolithic dome resembling the existing Beggs Event Center (shown here) will house Beggs School District's tornado safe room. (Photo source: Monolithic Dome Institute)

maintenance budgets. If a disaster occurs, insurance and grant funding may be used to rebuild in a more resilient manner.

In order to garner the needed political and financial support, it is critical that as many key stakeholders as possible be involved in these efforts. This builds understanding of the need for this work and support for obtaining the needed funds and frequently results in better decisions that meet the needs of the entire community. In particular, school leaders should connect with other leaders including elected officials, members of the local business community, and other trusted leaders who have a stake in having functioning schools in the community. Some school leaders have effectively involved students in these outreach efforts, whether it be giving high school seniors over age 18 and staff time off to go vote on relevant school safety initiatives, or engaging younger children in rallies and community meetings to draw attention to the cause. See Chapter 6 for more information on community outreach.

3.5   Importance of Quality Assurance Measures

3.5.1   Overview of Design and Construction

Peer review, in which one engineer or architect checks the work of another, is often a good practice that can help ensure that the intended design goals are achieved. School buildings should be constructed by highly qualified professionals, using appropriate materials, according to the approved design. Additionally, rigorous oversight of construction improves the ultimate quality of the building.

Design review to ensure compliance with the building code, as well as oversight and inspection of construction to assure compliance with the plans are known to greatly improve building performance in natural hazards. For example, it has been demonstrated in actual earthquakes that buildings that have had their designs reviewed for compliance with the code have resulted in superior performance (CSSC, 2007). Similarly, damage to the Donald T. Shields Elementary School in Texas due to a December 26, 2015 tornado was due to wall connections that did not comply with the plans and specifications (Thompson, 2016). Additionally, a 2009 report to the California Seismic Safety Commission concluded that, due to thorough plan review and construction inspection, California schools have consistently out-performed other similar buildings in earthquakes (CSSC, 2007).

3.5.2   Long-Term Maintenance and Improvements

Ongoing attention to buildings and resilience to natural hazards is needed to ensure that schools remain safe years after their construction. Actions taken

---

FEMA-P-1000   **3: Making School Buildings Safer**   **3-13**

to retrofit and improve the safety of facilities may be carried out over a number of years and require tracking to ensure the program is completed. The knowledge of hazards can change with time and as events occur, which may require revisions to strategies for hazard resilience. For example, seismic retrofits conducted in the 1970s and 1980s may no longer be considered adequate today. Engineering consensus on design criteria for tornado safe rooms are very recent and is likely to evolve. Finally, ongoing maintenance and replacement of deteriorated building components is critical, to make sure buildings retain the hazard-resistant characteristics with which they were designed.

Chapter 4
Planning the Response

When an emergency happens during school hours or a school event, school personnel must assume the role of emergency responders for the children in their care because they might be the only ones present at that moment. Professional first responders may be delayed due to excessive demands for service or to damaged infrastructure or other conditions that impede response, such as roads blocked by rubble and debris, which can occur due to a widespread disaster. School responders must remain flexible in their approach, and able to cope with the dynamic disaster environment. Developing an effective Emergency Operations Plan (EOP) gives school personnel, including administrators, teachers, and staff, definitive direction for what to do before, during, and after an emergency. It provides clear and actionable direction on the "who, what, when, where, why, and how" for emergency response.

Because creative problem solving is critical to successful disaster response, a EOP's structure, process, and procedures need to allow for flexibility and improvisation. Every scenario cannot be planned out in advance. If the right leaders and partners contribute to the development of the EOP and work as a team using adaptive approaches that support a standardized process, the school is more likely to function well in the event of an emergency. Although having the EOP is important, equally important is having stakeholder networks, connections, and organizational capacity to implement the plan. Combined, these elements lead to effective response and resilience.

The purpose of this chapter is to introduce the issues that schools need to understand and consider for creating, maintaining, and practicing the use of an EOP for effectively responding to natural hazard events. In particular, this chapter covers the following:

- An overview of the purpose of an EOP;

- A description of the recommended process to develop an EOP;

- An overview of the structure and content of an EOP;

- Legislative aspects that should be considered when developing an EOP;

- A description of the importance of training and exercises; and

- Guidance on making an EOP actionable.

There are many resources that can help school personnel put together complete and effective EOPs, some of which are referenced within this chapter and others within the *Resources Appendix*. In particular, this chapter builds upon the *Guide for Developing High-Quality School Emergency Operations Plans Guide* (U.S. Department of Education, 2013), which provides the latest guidance on developing school EOPs that is applicable to many hazards.

Planning Principles

The *School EOP Guide* (U.S. Department of Education, 2013) identifies the following as key principles in developing a comprehensive school EOP:

Planning must be supported by leadership. Strong support for the planning team by local leaders, including district managers and senior-level officials, can help develop plans that are more effective.

Planning uses assessment to customize plans to the building level. Comprehensive, ongoing assessment of the school community is important for developing plans that are appropriate for the particular context, situation, and circumstances.

Planning considers all threats and hazards. It is important to consider all threats and hazards that might impact the school when developing a comprehensive EOP. The EOP should consider the needs before, during, and after these potential incidents.

Planning provides for the access and functional needs of the whole school community. Planning should be inclusive of all the school community, including those with access and functional needs, those from diverse backgrounds, and those with limited English proficiency.

Planning considers all setting and all times. When developing an EOP, it is important to consider events that could happen during and outside of school hours, as well as in or outside the school premises.

Creating and revising a model EOP is done through a collaborative process. Using a collaborative process to create and revise an EOP helps to ensure that the plan is inclusive and effective.

More information on these principles can be found at: http://rems.ed.gov/K12PlanningPrinciples.aspx.

4.1 Purpose of a School Emergency Operations Plan

Thorough response planning for a hazard event helps ensure that a school community executes an organized, timely, and well-communicated response when the unexpected occurs. Every school should develop and maintain an EOP that clearly states what actions need to be taken before, during, and after an emergency event, who is responsible for those actions, and contingencies for the different situations that could arise. The plan should provide enough details that it can be actionable, easily understood, and readily used.

If any part of the school campus will be used as a shelter during or following a natural hazard event, special considerations should be included in the EOP. For example, plans for storing or procuring adequate amounts of water, food, and medicine for occupants should be established, as well as plans for providing adequate power supply (e.g., backup generators and storage of sufficient fuel supply). Parties responsible for developing the plans and ensuring that they are carried out will vary by context and should be identified. Alternate responsible parties should also be identified.

Administrators, teachers, and staff need to know their roles and be trained accordingly. The entire school community needs to practice responding to an event so that everyone reacts appropriately when a disaster occurs. School leaders should know how to interface with community partners, such as local fire, police, and other emergency personnel. EOPs should also incorporate school preparedness and mitigation strategies and activities, and specify up-to-date school safety policies and protocols.

An EOP is a good way for school personnel to think through and be ready for all of the difficult issues that emergency events bring. It also protects financial investments and helps build a culture of personal safety in the school community.

4.2   Recommended Process to Develop an EOP

EOPs are best developed by a diverse team of school stakeholders with a range of knowledge and perspectives. This team may include participation of the administration, teachers, staff, students, parents, and representatives of school community members with disabilities and minority groups. It also includes school support staff who may not be always be on campus, such as bus drivers and substitute teachers. Involving students in developing the school EOP (and overall safety preparedness) builds student leadership, engages youth in school emergency management planning and promotes student preparedness.

> EOP ASSIST 2.0, the Readiness and Emergency Management for Schools Technical Assistance Center's software application, is a valuable resource that is designed to help schools create and update high-quality emergency operations plans. More information at this link: https://rems.ed.gov/EOPASSIST/EOPASSIST.aspx

School safety planning goes beyond the school campus. It involves critical collaboration with local emergency responders, businesses, community groups, parents, and other key stakeholders with whom the school might need to interact during an emergency event. For example, a nearby business whose parking lot will be used for evacuation is an important planning team member. For more detailed information on involving the community in developing emergency operations, see Chapter 6.

In particular, the *School EOP Guide* recommends the following six-step process, shown in Figure 4-1, to develop, review, approve, and maintain emergency plans.

The National Incident Management System (NIMS) provides a common, nationwide approach for managing emergencies. Although it is not required for schools to use NIMS (unless receiving federal preparedness funding), the U.S. Department of Education recommends that all schools implement NIMS. NIMS provides the vocabulary, systems, and processes to enable schools to manage their emergencies and more effectively coordinate with their community's first responders

> Extensive online resources are available to help school administrators learn more about NIMS and ICS. See the *Resources Appendix*.

One element of NIMS is the Incident Command System (ICS), a nationally recognized management system that includes a common organizational structure, defined roles and responsibilities, and standard procedures. ICS has become the standard for emergency management across the country. ICS also provides school leaders with a management system that has been proven effective across incidents of all types and scale. Using ICS assists school administration and staff to coordinate effectively with each other and with first responders. It provides a common organizational structure, and allows school personnel to communicate with first responders using the same terminology and expectations, which leads to better alignment between agencies and fewer misunderstandings during an emergency. See Figure 4-2 for an example of how one school district adapted ICS to meet their needs.

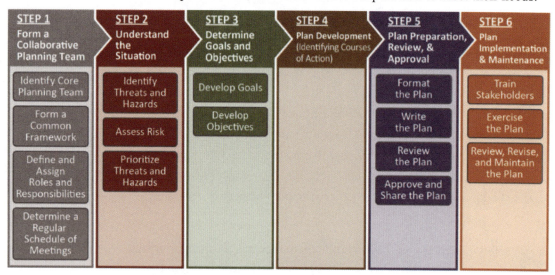

Figure 4-1　Six step process to develop, review, approve, and maintain a school Emergency Operations Plan (U.S. Department of Education, 2013).

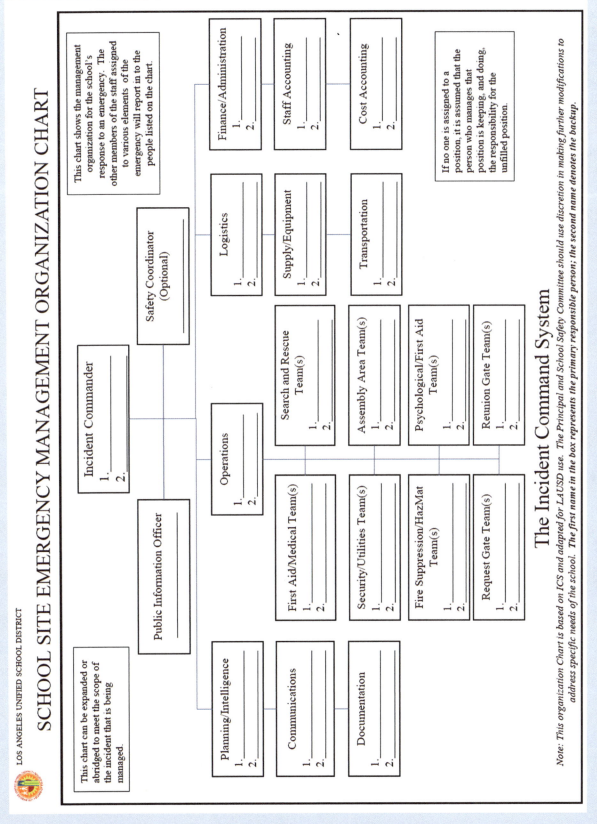

Figure 4-2　Example Incident Command System chart. (Source: Los Angeles Unified School District)

> The U.S. Department of Education's Readiness and Emergency Management in Schools Technical Assistance Center provides courses on EOP development by request. Visit the following website for more information: http://rems.ed.gov/ For other resources to support teacher training, see the *Resources Appendix*.

4.3   Overview of Structure and Content of an EOP

Developing and writing an EOP fall under Steps 4 and 5 in the process illustrated in Figure 4-1. School EOPs should be user-friendly and should be agreed upon by all parties that will play a role in the EOP. The specific structure of the school EOP can vary—one that works best for a particular school and context should be selected. Figure 4-3 provides an example of one EOP structure. This particular structure is often referred to as the "traditional format" and is provided as an example in the *School EOP Guide*. This structure, in particular, is described as follows and includes: (1) the Basic Plan; (2) Functional Annexes; and (3) Threat- or Hazard-Specific Annexes.

Basic Plan

1. Introductory Material	2.4. Planning Assumptions
1.1. Promulgation Document and Signatures	3. Concept of Operations
1.2. Approval and Implementation	4. Organization and Assignment of Responsibilities
1.3. Record and Changes	5. Direction, Control, and Coordination
1.4. Record of Distribution	6. Information Collection, Analysis, and Dissemination
1.5. Table of Contents	7. Training and Exercises
2. Purpose, Scope, Situation Overview, and Assumptions	8. Administration, Finance, and Logistics
2.1. Purpose	9. Plan Development and Maintenance
2.2. Scope	10. Authorities and References
2.3. Situation Overview	

Functional Annexes
NOTE: This is not a complete list, but it is recommended that all EOPs include at least the following functional annexes:

1. Communications	6. Reunification
2. Evacuation	7. Continuity of Operations (COOP)
3. Shelter-in-Place	8. Security
4. Lockdown	9. Recovery
5. Accounting for All Persons	10. Health and Medical

Threat- or Hazard-Specific Annexes
NOTE: This is not a complete list. Each school's annexes will vary based on its hazard analysis.

1. Hurricane or Severe Storm	5. Mass Casualty Incident
2. Earthquake	6. *Active Shooter*
3. Tornado	7. Pandemic or Disease Outbreak
4. Hazardous Materials Incident	

Figure 4-3   Example structure of a school Emergency Operations Plan using a traditional format (U.S. Department of Education, 2013).

4.3.1   The Basic Plan

This section addresses all of the activities a school must conduct during an emergency, regardless of the situation. This includes describing how the plan gets activated, assigning emergency responsibilities, defining the decision-making process, outlining administrative processes and controls, and explaining how the plan will be practiced and maintained over time.

Plans need to cover all times and locations that could impact school activities. For example, what if a natural hazard event occurs right before the school day begins or ends, during a school field trip while buses are transporting students, during an evening school event, or during a school-sponsored sporting event? Students may be unaccompanied on campus when an event occurs. Further, school personnel should be aware that they will likely be responsible for children after hazard events until they are reunited with their parents or guardians, which can take some time given the potential damage and disruption that can be caused by hazard events.

> An EOP should consider and address the access and functional needs that schoolchildren and staff might have, as well as injuries they might receive. These might include needs associated with limited mobility, visual and hearing impairments, cognitive delays, developmental disabilities, and multiple disabilities.

Specific requirements of a school's unique population should be addressed in the development of the EOP. Individuals with disabilities and those with access and functional needs may need assistance with language interpretation, special transportation and immediate medical needs. Resources to address these needs are available in the *Resources Appendix*. Individuals unfamiliar or inexperienced with local natural hazards or disaster response activities (e.g., transfer students, recent immigrants) may benefit from customized information on locally relevant hazards and basic disaster response activities. Additionally, if the school is designated to become a community shelter, this factor must be accounted for during the relocation of the students, guests, and staff.

4.3.2 Functional Annexes

Functional Annexes describe critical operational functions and courses of action that could be triggered by a variety of emergency events. While Functional Annexes address overall response operations in general, details may vary depending on the specific hazard. For example, evacuation sites and routes for a fire in the school building might be different than those needed for a tsunami. The following lists Functional Annexes that are particularly relevant to natural hazards, and provides examples of response operations for natural hazard events.

> An EOP should take into account forms of diversity that could affect planning activities. Factors to consider include ethnic background, non-English speakers, number of children on free and reduced lunch, transportation dependence, and other indicators that may influence ability to prepare, respond, or recover among students and staff. These considerations are becoming increasingly important as U.S. student demographics are rapidly changing and becoming more diverse and lower-income than the U.S. population as a whole (Southern Education Foundation, 2015).

Communications Annex. During and after a disaster, it is important to communicate with all stakeholders in a timely manner, providing what is legally appropriate and necessary, in a rapid, truthful manner. Schools should have procedures for communicating with administration, staff, students, families, the community, and first responders before, during and after an emergency. Considerations for effectively communicating with students with access and functional needs should be included. It is important

FEMA P-1000   4: Planning the Response   4-7

to ensure that relevant staff members can operate communications equipment. Using multiple forms of communication increases the likelihood of reaching everyone even if some communications channels are not functioning after a hazard event. This annex should include specific guidance on warning and alert systems, as well as protocols and procedures from local first responders and other authorities. Schools should also be prepared for media and citizen journalists, members of the public who report information (usually on social media), covering an event. Journalists are likely to focus on issues of public safety and the safety of the students and school staff. Schools should think through how to communicate impacts on students to the community. More detailed guidance on crafting and distributing effective emergency communications can be found in Chapter 6.

Evacuation Annex. Students and visitors may need to be evacuated from the school buildings and grounds during or after an event. For natural hazards, it is important to recognize that pre-designated onsite or offsite evacuation site(s) via preplanned evacuation routes must be selected with hazard zones in mind and may need on-the-spot adjustment due to debris or inaccessibility caused by a natural hazard event. For example, following earthquake shaking, an adult supervisor should ensure that evacuation routes are safe and unblocked before commencing evacuation. In communities with a tsunami hazard, the evacuation plan must consider tsunami inundation areas. In general, it is recommended to develop a plan on what to do when the primary evacuation route or evacuation area is unsafe or unusable.

> Best available refuge areas refer to areas in an existing building that have been deemed by a qualified architect or engineer to likely offer the greatest safety for building occupants during a tornado. These areas are not specifically designed to withstand tornadoes and should be regarded as an interim measure only until a tornado safe room/shelter is made available.

Shelter-in-Place Annex. With safety and security being of utmost importance, schools need to balance the risk of changing locations with staying in place. If sheltering onsite is needed during a natural hazard event, shelter areas should be pre-designated and ensured to be as safe as possible. For example, best available refuge areas that are to be used during a tornado should be selected by a qualified architect or engineer. These sites or safe rooms for sheltering in place during a response should be accessible for all members of the school community and conducive to meeting the needs of the individuals with disabilities and those with access and functional needs. The sites should be maintained with proper supplies for life safety and medical needs for the possible duration of a hazard event.

Accounting for All Persons Annex. Procedures for accounting for the safety and well-being of all students, staff, and visitors must be flexible enough to work in any location. Likewise, plans should recognize that communication and technology networks, such as cell phones and email, could be out of service during and after a hazard event.

Reunification Annex. The reunification process should be well documented and clearly communicated to parents, guardians and students in advance—this produces a process that is effective, builds trust, and reduces fear. It is critical that parents and guardians know how and where they should reunite with their children after a hazard event. For example, in communities where schools are located in the tsunami inundation zone, parents should know to go to the evacuation site, not the school. The reunification site should be accessible for all children and parents, and alternate reunification sites should be pre-determined in case primary sites are unsafe or inaccessible. FEMA provides a list of reunification systems that may be available to the public during a disaster: https://www.fema.gov/how-do-i-find-my-family.

> "Nearly half of parents, 45%, do not know the location to which their child would be evacuated as part of their school's disaster plan... Slightly over half of U.S. parents surveyed believed the school buildings in their community could withstand a major natural disaster such as an earthquake or tornado. Even among those parents who trusted in the physical integrity of the school buildings, however, 61% would ignore evacuation orders and retrieve their children" (Redlener et al., 2008).

Continuity of Operations and Recovery Annexes. Schools need to address the anticipated recovery continuity of operations needs of their specific school populations by identifying the needed and available capabilities and resources for recovery, and describing how these resources will be coordinated and mobilized. These annexes should describe how essential functions, such as payroll and educational continuity, will continue during and following an event. Clear and concise procedures reduce the adverse impact of a disaster event and help the school recover as rapidly as possible. Chapter 5 addresses the issues and planning for continuing school operations and recovery in more detail.

Health and Medical Annex. Medical and public health issues can be very sensitive and anxiety-producing for staff and parents and thereby require careful communications and planning with key partners. Children and adults experience very personal reactions to emergencies. This annex should address the medical and mental health needs of both staff and students and should consider that first responders might be delayed. This annex should identify where emergency medical supplies should be located and who is responsible for maintaining the supplies. Staff should know school site triage and what types of first aid they can administer and what they should leave to medical professionals. They should also know whether, when, and how to move seriously injured individuals, to ensure they do not cause additional injuries while trying to help. Chapter 5 provides more details on mental health considerations.

4.3.3 Threat- or Hazard-Specific Annexes

Schools with natural hazard risk need to consider the types of damage that a natural disaster can cause as they develop annexes for their EOP. Each school should develop the Threat- or Hazard-Specific Annexes that pertain to their school's specific natural hazard risks. The hazard-specific supplements

---

FEMA P-1000   **4: Planning the Response**   **4-9**

of this *Guide* indicate aspects specific to earthquakes, floods, hurricanes, tornadoes, tsunamis, and windstorms that should be considered, when relevant, in developing EOP Hazard-Specific Annexes.

These annexes describe courses of action that are specific to particular threats and hazards, such as "Drop, Cover, and Hold On" for earthquake response. Schools should develop annexes for all of the threats and hazards that they could face, based on a hazards analysis. In many cases, Threat- or Hazard-Specific Annexes will point to Functional Annexes, such as the Evacuation Annex.

4.4 Legislative Considerations in Developing EOPs

There are local, state, and federal laws that impact emergency planning. During the planning process, it is important for schools to abide by the laws that regulate their state and community. Federal laws that are relevant include the following:

> "More students with increasingly serious disabilities are being placed on general education campuses because of Least Restrictive Environment requirements. Those students lose the benefits of a specialized setting with greater numbers of skilled adults. This fact may broaden the effects of a disaster to a wider range of schools" (Barnes, 2013).

- **Americans with Disabilities Act (ADA)** prohibits disability discrimination across the spectrum on all emergency services, programs, and activities. For example, plans must address the provision of aids and services to ensure effective communications, that individuals are not separated from their service animals and assistive devices, and that a person with disabilities can receive disability-related assistance throughout an emergency.

- **Title VI, Civil Rights Act of 1964** establishes language access for providing effective communications with individuals, including students and parents, with limited English proficiency.

- **Family Educational Rights and Privacy Act (FERPA)** protects the privacy of student records and the release of student information.

- **Health Insurance Portability and Accountability Act (HIPAA)** requires health care providers and organizations to develop and follow procedures that ensure the confidentiality and security of protected health information when it is transferred, received, handled, or shared. HIPAA may apply to some schools (www.hhs.gov/hipaa/).

For state-specific mandates, see the REMS infographic at http://rems.ed.gov /StateResources.aspx, which provides information by state. More information on these laws and others that might apply to specific situations is available in the *Resources Appendix*.

4-10   **4: Planning the Response**   **FEMA P-1000**

4.5   Training and Exercises

Regular training, drills, and exercises make a plan effective and keep it relevant. Training builds awareness and understanding of specific response protocols and increases the chance that plan procedures will be followed appropriately during a disaster. School training activities should include all key stakeholders—regular staff, new staff, and those who might not be on campus regularly, but serve important roles, such as bus drivers and substitute teachers. Everyone should be familiar with the school's response protocols and their expected roles during an emergency. Exercises and trainings related to disaster response can be integrated into and built upon existing school activities. Knowledge gained from practicing emergency response in a school carries over into the home and the community. Additionally, coordinating with local leaders and first responders on school exercises and drills opens the door for opportunities for schools to participate in community-level exercises.

> "Large-scale drills and simulations ... illuminate roles and responsibilities, arrangements and connections for the complex coordination of disaster response. Demonstrated proficiency in a simulation has proven to result in better preparedness in real life" (Risk RED, 2009).

A wide variety of opportunities are available for training teachers, who are the first line of communication with the students. For example, teacher in-service workshops or trainings are an opportunity for teachers to understand and engage in the development and implementation of an EOP. In addition to learning procedures and protocols, these workshops increase understanding of local hazards, risk, and threats.

Exercises and drills of an EOP are essential. Exercising a school's plan lets everyone become familiar with their designated emergency roles and helps identify gaps, weaknesses, and improvement opportunities. It tests the feasibility and safety of evacuation routes as well as secondary (back-up) escape routes along with all of the other response procedures and protocols. Each practice response provides opportunity for students and staff to provide input and feedback. First responders can be engaged to observe and evaluate. All drills and exercises should be conducted in "no fault" environment that emphasizes learning and improvement. The lessons learned from each practice helps the school community respond more effectively in the case of a real event and improves response planning and overall school preparedness.

There are a range of exercise types a school can conduct. In general, it is recommended to partner with emergency managers and to involve other key players in any exercise. Some examples include:

- **Tabletop Exercises.** These exercises consist of small group discussions that go through emergency scenarios and the actions that need to be taken before, during, and after an incident. These discussions can be

---

FEMA P-1000   4: Planning the Response   4-11

incorporated into curriculum and classroom activities, as well as safety-oriented school meetings or workshops for the school (including the parents).

- **Drills.** This is the most common way to exercise the school and test an EOP. Schools most often conduct drills throughout the academic school year. During a drill, the school personnel use actual school grounds to practice responding to a scenario. It is also important to consider and involve parents in drills to make sure they understand the process for reunification with their children if there is an emergency. In addition to the standard fire drill, drills are important for practicing responses to any emergency that might occur in a school. For example, Great ShakeOut Earthquake Drills are an annual opportunity for schools (and people in homes and organizations) in earthquake-prone areas to practice what to do during earthquakes and to improve their preparedness.

- **Functional Exercises.** These are similar to a drill, but involve multiple partners, perhaps district-wide, and typically employ the Incident Command System structure.

- **Full-Scale Exercises.** These are multi-agency, multi-jurisdictional "boots on the ground" drills that test assets such as communication systems and equipment. These exercises help build community resilience and focus on continuity of services following a disaster.

Response exercises are a critical foundation to school, community, and national preparedness efforts. Support for designing, executing, and evaluating all of these exercise types are offered in the Homeland Security Exercise and Evaluation Program (HSEEP). More information about HSEEP and planning and implementing exercises is available in the *Resources Appendix*.

4.6   Making the Plan Actionable

Once an EOP is developed it needs to be communicated and discussed with the school community and key stakeholders. The plan should allow for flexibility and improvisation when responding. It must be regularly reviewed and updated with changes to school policies and procedures, building and campus improvements, contact information, and response activities. Training and exercises are critical for making sure the plan works and is understood by all involved. All these elements combined will help ensure the safety of students and staff during an emergency.

Training for Severe Weather: Exercise Play in Oklahoma

Severe weather regularly occurs during the school year in Oklahoma, forcing school administrators to make informed decisions with limited lead-time. An interdisciplinary team comprising University of Oklahoma professors and a local emergency manager questioned how they could best prepare teachers, administrators, and emergency managers to work together to solve problems in an emergency. Building off a statewide survey and focus group concerning information needs and communications in severe weather, they created a tabletop exercise to increase stakeholders' knowledge and awareness of their responsibilities during severe weather. This problem-based learning activity was also designed to train them on the importance of proactive decision-making during a severe weather event, as well as to highlight the importance of communication among the different groups.

Participants in the tabletop exercise were selected from three Oklahoma school districts and included a public emergency manager, two school emergency managers, and a variety of school decision makers including superintendents, principals, teachers, coaches, transportation directors, and maintenance directors. Participants were assigned to groups based on their roles and affiliations for each of the problem-based activities. Simple index cards were used to trigger the play. A "Time Stamp" card used actual case data from the National Weather Service including alerts, outlooks, and damage reports. A "Happenings" card introduced school events or everyday actions, such as parents calling the school to check on band concerts. Each round of the game started with a "Critical Event" card triggering discussion about what had happened and how participants would respond. It posed questions such as: *What actions do you take? Why? How concerned are you? How confused are you and what is confusing you?* For most questions, participants responded using a scaled ranking from 1 to 10. Participants shared decisions, and deliberated and recorded their discussions. Afterwards, participants reflected on how they would respond to future events and noted the benefits of networking. One participant wrote afterwards, "Being able to connect with other districts to talk about their plans during severe weather was extremely beneficial." (Stalker et al., 2015)

Figure 4-4   Participants taking part in the tabletop exercise.

Chapter 5

Planning the Recovery

Recovery actions typically begin after the emergency response phase ends. While the emergency response period is typically brief—lasting only minutes, hours, days, or weeks—recovery may take months or even many years after a major event. Recovery typically refers to putting a disaster-stricken community or organization back together. In the case of schools, the goal often focuses on the restoration of education and learning, as well as recovery for the people who make up the school system.

Post-disaster recovery is often more challenging and time consuming than people expect. Thus, planning the recovery—which includes understanding the various steps, policies, persons, costs, and opportunities involved in this process—can greatly facilitate the speed and effectiveness of post-disaster recovery activities within schools and surrounding communities.

The purpose of this chapter is to provide an overview of the issues schools can face after a natural disaster and the steps school leaders can take to be ready to recover well. In particular, this chapter includes the following:

- Important aspects to consider while trying to get students and staff back into school buildings;

- Considerations related to the health, safety, and well-being of students and staff during recovery;

- Guidance on financing the recovery; and

- The importance of planning for the next event.

The following sections highlight considerations specific to post-disaster recovery. Information about recovery that is unique to particular natural hazards is presented in the hazard-specific supplements of this *Guide*.

5.1 Getting Back in School Buildings

5.1.1 Post-Disaster Building Assessment

After any natural disaster, a first step is expert assessment of whether school buildings are safe to re-enter and use. These assessments are generally conducted by local building officials or other trained experts. Depending on the type of natural hazard, different technical standards exist for these post-

> School leaders should reach out to local building professionals to pre-arrange school building evaluations after a disaster. Local building professionals will likely be very busy following a hazard event and might not be immediately available for inspection without prior arrangement.

FEMA P-1000   5: Planning the Recovery   5-1

event building safety assessments to ensure consistency and technical rigor. These standards are discussed for each hazard in the hazard-specific supplements of this *Guide*.

The Four Most Fundamental Kinds of Recovery

The *School EOP Guide* (U.S. Department of Education, 2013) identifies the following as fundamental aspects of recovery:

Academic Recovery. This should consider school opening/closings, alternate sites if the school building cannot be used, and alternate education if students cannot physically reconvene.

Physical Recovery. This entails documentation of school assets, records management, inspections, damage assessments, public access, and security.

Fiscal Recovery. This includes funding for recovery efforts, records, legal aspects, and donations management.

Psychological and Emotional Recovery. This includes mental health services, memorials, and event commemorations.

5.1.2 Documenting the Damage

It is important to document damage to buildings, contents, and equipment shortly after a disaster, but prior to any substantial clean up or repair activities. Documentation of damage can be important for many reasons, especially in terms of receiving appropriate insurance or other reimbursement for cleanup, repairs, and replacement of damaged items. This entails thorough photographing of all damage, as well as documentation of time and materials spent on the response and recovery. Ideally, documentation of damage is also conducted by registered building design professionals, who know which types of damage are of most concern. Figure 5-1 shows examples of photos documenting earthquake damage.

5.1.3 Building Back Better

After a disaster, but before students and staff can get back to work in school, buildings often need to be repaired, cleaned, and in some cases retrofitted or demolished and replaced. The timing and funding needed for this work can vary tremendously depending on the damage that occurs.

In some cases, a damaged building may be required to be repaired or rebuilt in compliance with the latest hazard regulations and building code requirements. The rules around what level of damage triggers these requirements vary significantly across jurisdictions—some communities may

Figure 5-1   Photos documenting earthquake damage to school buildings. (Photo source: Michael Mahoney, FEMA)

Safer, Smarter, Stronger… Greener

On May 4, 2007, an EF5 tornado tore through the small town of Greensburg, Kansas. The tornado was more than a mile wide—wider than the town itself. It ultimately destroyed 95% of the structures in the community.

As Greensburg dug itself out from the rubble, leaders made a commitment to build back better and more environmentally sustainable than before. After much planning and with the help of knowledgeable experts, Greensburg made a commitment to becoming the greenest town in the United States. Now, the city hall, the hospital, and the local school have all been built to the highest Leadership in Energy and Environmental Design (LEED) certification level offered by the U.S. Green Building Council.

The new school in Greensburg was "built green" from the ground up. School leaders worked with a professional design and construction team to plan, design, and build an environmentally responsible, student-focused academic environment that reinforces Greensburg's community-wide commitment to sustainability (U.S. Department of Energy, 2011).

---

have lower thresholds than others that trigger these requirements. If school buildings are also historic structures, they may be subject to a different level of thresholds that trigger these requirements.

These rules are known as "compliance triggers" and are typically expressed in terms of building damage, such as "substantial damage" or "substantial structural damage," both of which are defined in the *International Building Code* (ICC, 2014b). School leadership should check with their local building code officials about substantial damage and improvement rules that are relevant to them.

> FEMA developed a Substantial Damage Estimator (SDE) to assist state and local officials in determining level of substantial damage for buildings in flood hazard areas. For more information see: www.fema.gov/media-library/assets/documents/18692.

For schools that need to conduct major repairs or demolish and rebuild one or more school buildings, this is an unparalleled opportunity to build back better. During recovery, community members and government agencies will be keenly aware of the importance and consequences of natural hazards. With the right information, advocacy, and community-level support, they may be willing to support disaster-resilient building design. Designing and constructing new disaster-resilient school buildings is discussed in Chapter 3.

5.1.4   Adaptability

> After the 2016 floods in Louisiana, the Louisiana Connections Academy allowed displaced students to stay current in their school work through an online system that had become established in 2011 (Connections Academy, 2016).

School leaders should make contingency plans in case they cannot reoccupy some or all of their buildings in a reasonable timeframe due to the need for retrofit or reconstruction. These plans can range from using modular classrooms, to sharing space temporarily on another school campus, to occupying another building that is not a school, to online systems that facilitate educational continuity. These plans must be made in discussion with local and/or state building officials because many states have particular laws that regulate school buildings. These plans should reflect the likelihood of building damage based on pre-event hazard vulnerability evaluations of existing school buildings, as discussed in Chapter 3. It is critical that school leaders plan for these back-up facilities or systems, as this is what can ensure educational continuity for students. When back-up locations have been identified before an event, teaching and learning can often continue, even after a disaster that causes widespread damage and disruption.

5.1.5   Schools as Emergency or Recovery Shelters

> For schools designated by the local community emergency authorities as a shelter (known as Risk Category IV in the *International Building Code*), the required design loading is stronger than typical, intended to greatly increase the chances of the building being available as a shelter after an earthquake or hurricane event.

The role that schools can play in re-establishing routine can be complicated by the fact that many communities designate schools as shelter sites. Often, schools are one of the only public buildings with large assembly spaces, such as gyms, auditoriums, and cafeterias. Accordingly, schools can and often do serve a variety of shelter functions during the emergency and early recovery periods following disaster. In some cases, schools are a site that community residents evacuate to during a disaster, such as during a tornado or a tsunami. Schools often serve as post-event mass shelters, temporarily housing residents who cannot stay in their homes. Schools may also be designated as recovery centers, hosting a variety of post-disaster support services and personnel for the community. All of these functions are critically important. But they also can interfere with plans to resume school and to get students and teachers back in the classroom.

When Floods Displace a School: The Value of Planning Ahead

Six months prior to the 2013 Colorado Front Range flood, the St. Vrain School District updated their Continuity of Operations Plan (COOP) to identify a secondary location for students if they ever had to close a school because of an emergency. During this process, they held table top exercises involving directors from each of the major divisions of the district including emergency management, nutrition services, technology, custodial services, operations and maintenance, and human services. They developed two options for reunification if an emergency were to occur: (1) they would divide students between other schools across the school district; or (2) they would institute an agreement with the tenants of another nearby school to use their facilities as a temporary school location during displacement. Little did administrators know at the time that this plan would need to be implemented only a few months later.

In September 2013, Colorado experienced multiple days of record-breaking rainfall and flash flooding that resulted in 10 deaths and the evacuation and forced relocation of approximately 18,000 residents. The small community of Lyons, Colorado, was devastated. Due to road and bridge failures, residents were trapped for 36 hours as they waited for National Guard assistance. Lyons experienced an almost complete loss of services including power, telephone, sewage, and potable water. Due to the damage incurred by the town, residents were evacuated and unable to return for at least six weeks.

Although Lyons Elementary and Lyons Middle/Senior High Schools escaped the flood damage, the buildings remained inaccessible to students and staff due to the massive destruction in other parts of the community. Over 700 students were displaced. While this disaster could have caused chaos for an unprepared school district, the St. Vrain Valley School District was ready to act.

Within one week of the flood—and with the feedback and collaboration of principals from both schools—the school district announced that they would resume classes beginning the week of September 23, just 9 days after the floodwaters ravaged Lyons. Both schools were reopened 11 miles east of Lyons in a neighboring community. The students, faculty, and staff met on their regular class schedule for ten weeks at the temporary location until they finally returned to their home schools on December 2, 2013. Without the foresight of planning implemented by the school district, these children—and the faculty and staff that support them—could have faced many negative academic and emotional consequences as a result of being displaced and separated from their familiar school environment. For the sake of these children, planning truly mattered (Tobin-Gurley, 2016).

Figure 5-2   Schoolchildren affected by the Colorado floods. (Photo source: Peggy Dyer, One Million Faces Project)

School administrators should check with local emergency managers, American Red Cross officials, or other shelter providers to determine whether and how local schools may be included in current emergency and recovery plans for their community. Some school districts may wish to negotiate formal agreements with shelter providers in order to address issues including access to facilities, liability for damages and loss of functionality, and other relevant topics. The State of Washington, for example, has enacted laws that enable the state to hold harmless or indemnify the owners of private facilities, including independent schools, that permit the use of private land or facilities for public evacuation. Local emergency managers may be familiar with similar laws or statutes in other jurisdictions.

School district leaders and staff should consider contingencies for continuing instruction during emergencies that may require public use of their facilities. Once contingency plans are established, it is important to practice how these plans would be carried out during the school's disaster exercises and trainings. See Chapter 4 for more information on practicing plans. Also see the hazard-specific supplements for more information on evacuation or emergency shelters related to specific natural hazards.

5.2   Focusing on Routine and Mental Health

In addition to restoring physical infrastructure, it is also essential that school leaders focus on the health, safety, and well-being of students and staff during the recovery period. Although planning for this is often overlooked, it is important that leaders keep in mind how long this process can take and what a central role schools can play in facilitating individual, family, and community recovery. In addition, a major lesson of past disasters has been that people can move in and out of vulnerable conditions, and that recovery is not simply one linear, straightforward path for all. This means that human and financial resources may need to be dedicated over a long-term horizon— and targeted at different time points—to ensure that students and staff receive the proper support they need to fully recover.

5.2.1   Re-Establishing Routine

Disaster experts have long recognized that getting children back into a routine is a key driver of recovery. Schools can help provide the normalcy that children often desire, while also allowing adult caregivers to focus on re-establishing their own "new normal."

Hurricane Katrina: Restoring Routine

Hurricane Katrina made landfall at the beginning of the 2005-2006 school year. According to the U.S. Department of Education, approximately 372,000 students were displaced from their home communities in the states directly affected by Katrina, while 160,000 remained dislocated for years after the storm. In many cases, these children moved numerous times after the initial displacement, and each transition meant adjusting to a new home, neighborhood, school, teachers, and peers.

Re-establishing routine was justifiably a central concern for teachers and school administrators in both disaster-affected and receiving communities. Teachers reported that when the initial novelty of the "extended vacation" wore off for children, they actually seemed to want a routine in the classroom. Of course, given the enormity of the disruption caused by Katrina, school staff had to find the right balance between regimented school routines and the desire to be flexible and adaptable to support children's emergent and ongoing needs.

When classes resumed after Katrina, teachers across the Gulf Coast and beyond worked hard to maintain rules in the classrooms, although there was some recognition that exceptions to the normal rules would have to be made. For example, one student was not wearing the required knee socks with her school uniform. When the teacher pulled the student aside to reprimand her, she found out that the student had lost all of her clothing in the flood, and that she no longer had the appropriate attire to wear to school (Fothergill and Peek, 2015).

Figure 5-3   Drawing by child affected by Hurricane Katrina (Fothergill and Peek, 2015).

5.2.2 Assessing Mental Health

Until a new routine is established, and even afterwards, those impacted by disaster may experience issues with safety, stress, and extended grieving. Emotions can include hopelessness, depression, guilt, and withdrawal. In the most severe cases, classroom behavior and academic performance may be

> Psychological First Aid for Schools is an evidence-informed intervention approach to assist children and staff after a disaster, including natural hazard events. See here for information: www .nctsn.org/content/psychological -first-aid-schoolspfa. See the *Resources Appendix* for other post-disaster mental health resources.

impacted. Teachers and staff—who are coping with their own recovery while also managing school needs—also may be challenged by changed emotions and behaviors. Teachers and staff who were not affected by the disaster might also be faced with similar challenges given that in some cases, affected children are relocated to other schools after a disaster.

In the immediate aftermath of an emergency, students, staff, and families may need psychological support to help deal with trauma-related distress, which can a have a long-term impact. The National Child Traumatic Stress Network (NCTSN) has assembled resources for many forms of trauma experienced by children, including natural disasters. These resources include guidance about things teachers can do to help their students recover, and steps that teachers can take to help themselves cope. School crisis intervention teams can also play an important part of the recovery process.

Over the longer term, teachers often adopt new age appropriate arts- or writing-based curriculum to help children and youth process their disaster experience and to reflect on a new future in the aftermath of the event. Peer listening teams can also be powerful during the recovery period, in terms of engaging young people to become a part of their own and each other's healing process.

5.3   Financing the Recovery

Following a natural disaster, it is critical for schools to prioritize detailed legal and financial recordkeeping. These are needed for reimbursements of any expenses, including repair and reconstruction costs. They also might be needed for legal reasons.

Of course, these records are often needed at a time when accessing databases and files can be challenging or impossible, and all staff members are extremely busy, both with emergency responsibilities and with concerns for their own loved ones. This is why planning in advance to ensure that critical records are regularly backed up and stored offsite is invaluable.

After a disaster, concerned people might seek to give donations for response and recovery. Knowing whether and how schools can accept donations and developing systems to accept them, perhaps through partner community organizations, allows schools to benefit from this generosity. At the same time, donations management can become its own full-time job, as schools can quickly become overwhelmed as material goods—both needed and unneeded—arrive rapidly after a disaster. In some cases, schools have had to hire donations management experts or rely on volunteers to help process the influx of goods that arrive after a disaster. It is important for school officials

5-8   5: Planning the Recovery   FEMA P-1000 to clearly articulate what they do, and do not, need after a disaster, and to try to ensure that when donations are made, they are helpful to school operations and to student success.

5.3.1 Federal Resources

Federal resources can also assist in the recovery. In particular, the following are resources that might be relevant:

- The Stafford Act authorizes the delivery of federal technical, financial, logistical, and other assistance to states and localities during declared major disasters or emergencies. FEMA coordinates administration of federal disaster relief resources and assistance to states if an event is beyond the combined response capabilities of state and local governments.

- FEMA's Public Assistance (PA) program provides supplemental federal disaster grant assistance for debris removal, emergency protective measures, and the repair, replacement, or restoration of disaster-damaged, publicly-owned facilities or certain private not-for-profit institutions following major disasters or emergencies declared by the President. The federal share of assistance is not less than 75% of the eligible cost. The recipient (usually the state) determines how the non-federal share (up to 25%) is split with the sub-recipients (eligible applicants). The PA program also provides assistance for hazard mitigation measures during the recovery process. More information may be found on FEMA's website: http://www.fema.gov/public-assistance -policy-and-guidance.

> Following the August 2016 floods in Louisiana, FEMA's PA program awarded nearly $79 million to schools to support recovery efforts, including repair and rebuilding (FEMA, 2017a).

- FEMA's Hazard Mitigation Grant Program (HMGP) may provide federal funds for structural improvements after a disaster strikes. More information can be found here: http://www.fema.gov/hazard-mitigation -grant-program.

- Certain private schools might not be eligible for FEMA's PA program or HMGP. They may, however, be eligible for low-interest disaster loans from the U.S. Small Business Administration (SBA) to repair or replace certain property and equipment damaged during a declared disaster. More information can be found here: https://www.sba.gov/loans -grants/see-what-sba-offers/sba-loan-programs/disaster-loans.

Meeting Payroll: Nimble Financial Services Can Support Recovery

On Monday, May 20, 2013, a 2-mile-wide EF5 tornado struck the Oklahoma City suburb of Moore, killing 24 people, including 9 children, and causing an estimated $2 billion in property damage. Three public schools were especially hard-hit: Briarwood Elementary School, Plaza Towers Elementary School, and Highland East Junior High. The tornado rendered the Moore Public School District's administration building, including its technology center, unusable. Servers that contained payroll data for approximately 3,000 district employees were destroyed, making it impossible for the school system to make payroll. In accordance with their employment contract, teachers in the Moore District were due to be paid the Friday after the tornado. Proactive efforts by local banks and the State Banking Commissioner enabled the teachers and administrators to receive uninterrupted pay.

BancFirst, the account holder for the school district, worked with several local banks and state regulators to provide teachers in the district with provisional credit while the district worked to salvage its payroll data. The majority of financial institutions in Oklahoma decided to give Moore Public School employees interest-free, short-term loans for the amount of their last paycheck.

Oklahoma State Banking Commissioner Mick Thompson sent a public bulletin to state-chartered bank presidents on Thursday, May 23, asking the banks to provide provisional credit to district employees the next day, Friday, May 24. "As Oklahoma Bank Commissioner, I am asking that if you have customers/depositors who are employees of the Moore, Oklahoma public school system and who would be receiving a payroll deposit tomorrow, that you consider providing provisional credit until their payroll deposit arrives," Commissioner Thompson wrote.

"Basically what we told the banks is that the school district won't be able to transfer the money into your accounts, so just give the employees whatever the amount of their last payroll deposit was," Thompson said. (Bailey, 2013)

Figure 5-4   Briarwood Elementary was one of three public schools destroyed or damaged by the Moore, Oklahoma tornado of May 2013. (Photo source: Scott Olson, *Getty Images*)

5.4   Planning for the Next One

Even after full recovery from an event, it is important to dedicate time to evaluate the process and procedures used for emergency preparedness, response, and recovery. Plans including policies, procedures, roles, chain of command, and functional and hazard specific annexes should be examined to evaluate what worked and what did not. Plans should be revised to incorporate:

- Updates to all contact information for all key leaders, partners, and first responders;

- New and updated resource lists for long-standing and newly established contacts;

- Revisions to key response activities, such as evacuation routes and sites;

- Changes to the buildings, school campus, bus routes, and broader community;

- Changes to school policies and procedures;

- Lessons learned; and

- Any issues that may have been overlooked in the original plan.

Once the plans are updated, briefings should be held to communicate the plan modifications and improvements to key stakeholders, staff, and students, as well as media, parents, and local officials. This type of outreach encourages stakeholder engagement and provides an opportunity for dialogue with the wider community.

If at all possible, schools should also keep detailed notes regarding how the response and the recovery went during the actual event. This will ensure that important details can be recalled later when evaluating how plans and response activities could be improved. Some emergency officials invite researchers and social scientists trained to take systematic field notes to come and observe response and recovery activities.

Chapter 6

Engaging the Whole Community

School disaster resilience is most effectively achieved when the community is engaged in the process to understand and reduce school risks, plan for emergencies, and recover from damaging events. This chapter provides guidance on engaging the whole community and in particular, provides:

- An overview of the whole community approach and the role that schools play;

- Guidance on engaging community partners, including professionals, local government, community organizations, parents, and students, and why their partnership is important;

- Guidance on communicating with the community; and

- Information on tools and technologies for effective communication with the community.

6.1 The Whole Community Approach

For nearly a decade, FEMA has moved toward a "whole community" approach to emergency management. This approach recognizes that all resources and diverse segments of the community must be fully engaged in order to most effectively prepare for, protect against, respond to, recover from, and mitigate against all hazards. Partners in this work include FEMA and other federal agencies; local, tribal, state, and territorial leaders; schools and higher education; health care; non-governmental organizations including faith-based and non-profit groups; private sector industry; and individuals and families. The whole community approach recognizes and embraces diversity in its broadest definitional sense, including an array of organizational representatives, but also re-focusing efforts to ensure that members of historically marginalized communities (including the elderly, persons with disabilities, low-income populations, immigrants and non-English speakers, racial and ethnic minorities and others) are included in the discussions and that their voices are heard and respected.

At the heart of the whole community approach is the idea that all members of the community need to have a voice, as community members and leaders are

FEMA P-1000   **6: Engaging the Whole Community**   6-1 best situated to identify local assets, capacities, interests, needs, and goals. Put simply, they are the experts on their own local community and hence they should be engaged in the disaster planning process. The foundational principles of the approach, as described by FEMA, include:

- **Understand and meet the actual needs of the whole community.** Community engagement can lead to a deeper understanding of the unique and diverse needs of a population, including its demographics, values, norms, community structures, networks, and relationships. This will allow for better understanding of the community's real life safety and sustaining needs and their motivations to participate in emergency management-related activities prior to an event.

- **Engage and empower all parts of the community.** Engaging the whole community and empowering local action will better position stakeholders to plan for and meet the needs of the community. This type of engagement will also strengthen the local capacity to deal with the consequences of various threats and hazards. This requires all members of the community to be part of the emergency management team. When the community is engaged in an authentic dialogue, it becomes empowered to identify its needs and the existing resources that may be used to address them.

- **Strengthen what works well in communities on a daily basis.** A whole community approach to building community resilience requires finding ways to support and strengthen the institutions, assets, and networks that already work well in communities and are working to address issues that are important to community members on a daily basis. Existing structures and relationships that are present in the daily lives of individuals, families, businesses, and organizations before an incident occurs can be leveraged and empowered to act effectively during and after a disaster strikes.

Schools are an essential part of the whole community. Not only do schools educate future generations, they also serve as gathering places and hubs of community activity and vitality. Communities invest in schools through property taxes and other forms of support, and in turn, schools serve as sources of employment, hubs of leadership, and even symbols of hope for the future within communities.

The Societal Value of Schools

"Our society places great importance on the education system and its schools, and has a tremendous investment in current and future schools... The school is both a place of learning and an important community resource and center" (FEMA, 2010a).

There are over 50 million students attending close to 99,000 public elementary and secondary schools with an additional 5.2 million students attending close to 34,000 private schools (NCES, 2016). School facilities highly vary from one-room rural schoolhouses to citywide urban schools that have over 5,000 students.

An involved community increases support for needed funding, improves emergency planning by involving a wider range of voices, and makes response and recovery more efficient and effective because community members know what to expect and take steps to assist schools and school leaders. Not only does engaging a wide range of community stakeholders improve school safety preparedness, their engagement furthers a school's reach into the infrastructure of the community, improves institutional credibility, and expands a school's ability to call on an expanded pool of resources—human, financial, and material—when a disaster happens.

Engaging the community prior to a hazard event can help build a culture of preparedness that moves schools from a reactive response into a proactive planning mindset to create and sustain safety and security for the entire school community. School leaders often effectively interface with parents, nearby universities, and local emergency management; they should also consider reaching out to a broader swatch of potential partners, including community-based organizations, local business and industry, and government officials.

6.2 Engage Community Partners

After a disaster, a wide range of individuals, organizations, agencies, and disaster management professionals quickly rally to respond. Some turn to immediate life-safety response needs, and others focus on effective management of a school district's operational response and facility security. Those involved in the pre-disaster planning period often continue to serve active roles during response and throughout the recovery period.

> After a disaster is not the time to be exchanging business cards.

Because of the number of individuals and organizations involved throughout the disaster lifecycle, relationships need to be built and actively maintained over time. This can be achieved by involving local partners and community stakeholders in a variety of events that focus on preparation and education

FEMA P-1000   **6: Engaging the Whole Community**   6-3 regarding natural hazards. These may include commemorations of past damaging events or participation in drills. For example, the Great ShakeOut organizes a series of earthquake drills, which provide an annual opportunity for people in homes, schools, businesses, faith-based settings, and other organizations to practice what to do during earthquakes and to improve earthquake preparedness. These drills are carried out in many areas of the United States and around the world. Similarly, a number of communities practice community-wide tsunami evacuation drills, where schools partner with local government and community groups for a more realistic drill experience. Regardless of what the school decides to do—whether it be an annual event or a weekly gathering—this is the time to be creative in terms of designing outreach and relationship-building activities that make sense in the local community culture and can be sustained over the long-term.

There are many different stakeholders that school leaders should consider engaging. The prioritization of these different groups and entities may vary by school district. Information on potential partners are provided in the following subsections. Given that their level of importance will vary by community, they are listed in alphabetical order, and not in order of importance.

6.2.1   Children and Youth

It is critical to remember that children and youth are vital stakeholders in any school-based community preparedness, response, or recovery effort. Research has consistently shown that children regularly express the desire to learn about hazards and also to engage in activities that help adults, help other young people, and ultimately help themselves.

Schools could incorporate information about natural hazards, approaches to reduce society's risks to hazards, and specific natural hazard risk information relevant to a school and a community into school curricula in engaging ways. Raising awareness of youth and children, starting from kindergarten, about

---

A Lesson That Saved Lives

On the day after Christmas in 2004, ten-year-old Tilly Smith of Surrey, England, was on vacation in Thailand with her parents. Smith had learned the warning signs of tsunamis in her geography class at Danes Hill School just two weeks prior. Seeing the water recede from Maikhao Beach and recognizing that a tsunami could be imminent, she alerted her parents who helped to tell others and clear the beach. Smith's timely warning saved nearly a hundred tourists and local beachgoers. Maikhao was one of the few beaches on Phuket Island with no reported casualties during the Indian Ocean tsunami, the deadliest tsunami disaster in recorded history (The Telegraph, 2005).

natural hazards and how to respond to them can be life-saving. A number of resources are available for educators on this topic (see the *Resources Appendix*).

Recognizing that individual schools can be limited in the flexibility of their curricula, it is also important to consider programs to engage youth beyond school hours. Involving youth in safety preparedness and disaster response, both in the school and the community, builds student leadership and increases overall community resilience.

Youth Creating Disaster Recovery & Resilience

Youth Creating Disaster Recovery & Resilience (YCDR^2) is a research project for youth affected by disasters. YCDR^2 connected with youth in disaster-affected communities affected by disasters including wildfires, tornadoes, and floods in Canada and the United States. Young people have used art, video, music, and storytelling to discuss what they needed, the challenges they have faced, and how they might contribute to helping their friends, families, and communities recover from disasters. Figure 6-1 shows an example of art created following the 2011 Joplin tornado. More on creative youth stories can be found at: www.ycdr.org.

Figure 6-1   Young people in Joplin, Missouri shared their stories through the YCDR^2 project. (Photo source: Circular Motions Photography)

Students have a vital role in contributing to the safety of their school and community. Such work is central to their development as citizens and, some argue, is part of their ethical and moral responsibility to themselves, their peers, and school staff. Although there are legitimate and important safety limits that must be placed on student engagement in emergency management activities, schools should place a high priority on involving this critical stakeholder group in mitigation, preparedness, response, and recovery activities. If schools miss this opportunity, they not only deny students' unique learning and leadership opportunities, they also decrease their ability to harness one of their greatest resources available to them. The day of a major emergency, schools need all of the coordinated assistance they can receive; adequately trained, empowered students will step up to this role if they are given the tools and trust of adults to do so.

One example program to engage youth in natural hazard planning is to create a local Teen Community Emergency Response Team (Teen CERT). Schools may use Teen CERT to engage youth in school emergency management planning and promote student preparedness. Adapted from the adult CERT program, FEMA established Teen CERT to equip high school students with basic response skills and an understanding of emergency preparedness concepts. CERT Basic Training includes emergency preparedness and disaster response skills such as fire safety, light search and rescue, team organization, and disaster medical operations. More information is available in the *Resources Appendix*.

Teen CERT Builds Health and Safety Awareness

"The staff at Valley Mills ISD in Valley Mills, Texas felt compelled to begin Teen CERT training at the school due to the health and safety needs of the faculty, staff, and students. They saw a high level of need at the school due to those suffering from health issues like diabetes, asthma, seizures, and other physical injuries. CERT-trained students are aware of the different medical issues and are able to assist in the event of an emergency" (FEMA, 2012a). Figure 6-2 shows an example of the value of this type of program for Teen CERT members.

Figure 6-2   A member of Teen CERT explains why the experience has been so valuable to him. (Photo source: Laura Conway)

Additional examples of ways to engage students in the process to develop a school hazard safety strategy include assemblies, after-school activities, student clubs, school exhibits, competitions, and safety drills.

6.2.2   Design Professionals

Design professionals, such as architects and engineers, can be valuable partners in helping to keep students and school staff safe. They can help determine a school building's pre-event vulnerabilities and provide recommendations for improving the safety of the school, such as retrofit options. For new school construction, they can also provide guidance on decisions affecting environmental, health, and safety aspects. Design professionals also serve a critical role in assessing the safety and severity of building damage following a hazard event.

6.2.3   Educational Professionals

The wide range of professionals working in education can offer resources and guidance on policy, stakeholder relationships, continuity of academic programs, and much more. These individuals can be helpful in crafting school emergency management plans and working through the preparedness planning process. These professionals may include school district leadership teams, school board members, school administrators and staff, teachers, members of the school safety team and school crisis team, agency-appointed mental health agents, grief counselors, risk managers/legal counsel, and educational spokespersons.

6.2.4   Elected Officials

Elected leaders within a community often run on platforms associated with schools and education. This is true for school board members, but also for other elected officials such as the mayor, city manager, legislators, city council members, and county commissioners. These individuals often care deeply about the quality and functionality of local schools, and can actively engage with school system representatives to build awareness and support for planning, preparedness, response, and recovery activities. These elected officials are often called upon to speak on behalf of their constituencies, making a close partnership with school leaders all the more important.

6.2.5   Emergency Management Professionals

Local emergency management professionals can provide invaluable information to schools regarding state-of-the-art preparedness, response, and mitigation activities. School leaders should be on a first name basis with leaders from local fire departments, law enforcement agencies, emergency medical services, and emergency management, as these are the front-line responders who may provide lifesaving first response and can also help the school through the recovery process.

6.2.6 Labor Bargaining Units

National unions, such as the National Educational Association (NEA) and the American Federation of Teachers (AFT), as well as state and local unions, can help guide school districts on work-related and human resource issues during the planning process and help in recovery by spearheading drives to gather relief fund donations and offer disaster relief grants for union members.

American Federation of Teachers Disaster Relief Fund

"Help us show our union family in Louisiana that we have their backs. Donate today."

The call-to-action posted on the donations page of the American Federation of Teachers (AFT) website (www.aft.org/disaster-relief-fund) urged union members throughout the country to donate financially to help union members hit by the massive 2016 flooding disaster in Louisiana. AFT, an affiliate of the American Federation of Labor and Congress of Industrial Organizations (AFL-CIO) representing 1.6 million members, actively assists members affected by disaster events by raising funds through the AFT Disaster Relief Fund. Donation pages on websites and social media feeds of the national AFT organization, state federations, and local affiliates gather funds to provide direct relief. Disaster assistance efforts similar to those for Louisiana in 2016 have provided assistance to union members after other major events including Superstorm Sandy.

6.2.7 Local Business and Industry

Members of the private sector often donate generously to school efforts, whether it be funding high school sports teams or providing free food at elementary school band concerts. Cultivating these relationships can help secure contributions for mitigation, response, or recovery costs.

6.2.8 Local Community Organizations

The wide range of community-based and faith-based organizations that focus on social service provision and the collective good can often play an active role in identifying resources and supporting school and local community preparedness outreach activities. Such organizations include the American Red Cross Disaster Action Teams (DAT), Community Emergency Response Teams (CERT), Salvation Army, and Voluntary Agencies Active in Disaster (VOAD). See *Resources Appendix* for more information and links.

6.2.9 Local Hospitals

Hospital administration, staff, and primary care clinicians are vital players in assisting in a coordinated and integrated response and recovery, particularly during large-scale emergencies. During preparedness planning activities, they can provide school planning teams valuable guidance on healthcare risks and prevention planning, as well as collaborate on medical response protocols and procedures (e.g., triage) that work most effectively for that school.

6.2.10 Local Jurisdiction Public Agencies

Departments, such as public works, planning, public health, healthcare, and family support services, provide specialized services to their jurisdictions and key community facilities and populations. These local departments and agencies are often natural partners for schools, as they are often focused broadly on the health, vitality, and sustainability of local communities.

6.2.11 Media

It is helpful to develop solid relationships and strategic partnerships with key media representatives and online social media influencers prior to an event. Journalists and other members of the media—including traditional print and broadcast media and social media influencers (individuals and organizations who are influential online)—have enormous power in terms of informing the public of pressing issues and ultimately shaping local political and social agendas.

During times of disaster, media often interface between school administrators and staff, students, families, and members of the community. As such, building relationships with media and online influencers to help cover key issues associated with schools in a thorough and thoughtful manner is crucial. In today's environment, which has been marked by an unprecedented number of school shootings and rising disaster risk, school officials should be prepared for local (and perhaps even national) media to take an interest in school safety issues. When an event occurs, media regularly spotlight the school's response. Social media channels trend heavily with posts and messages focused on topics and issues critical to the disaster event, response, and recovery. Media professionals are not experts in natural hazards and may be limited in their understanding of the events. When questions are asked, they expect the truth, including things which the school may not yet know. Timely and regular release of clear and consistent information should to be provided to the formal press and all interested individuals who may be covering the events.

Ideally, all this information should come from a single point of contact, usually a designated Public Information Officer of the school (or district) or a lead school official. Access to school spokespersons, leaders, and other subject matter experts should be provided, though access to impacted school facilities, staff, and particularly students should be closely regulated. Children—and the release of information regarding specific details on impacts to children—need to be protected for ethical and legal reasons (e.g., if a child has been affected, parents should not first find out through the media). Additional resources for developing messages and working with the media can be found in the *Resources Appendix*.

6.2.12 Parents and Caregivers

Parents, grandparents, and other caregivers are key stakeholders in the children's lives. Parents and others invested in children's futures provide a wide range of expertise, educational awareness support, and resources that can be tapped for preparedness activities. Parents can also serve as champions for enforcing safety at schools through legislative channels.

Parents as School Safety Champions

Much of Oregon is at high seismic risk, and some of the most vulnerable buildings in the state are schools. In response, Portland parents have mobilized to advocate for seismic improvements in schools.

Parents at Sunnyside Environmental K-8 in Portland, for example, formed a "Rock and Roll Committee" to draw attention to seismic issues. The group began by researching the structural integrity of the school building, which was built in 1925. A rapid visual screening of schools by Oregon's Department of Geology and Mineral Industries deemed Sunnyside as one of dozens of schools at "very high risk" of collapse in a strong earthquake. The Rock and Roll Committee hosted a dinner event to educate fellow parents and community members about the risks facing their school, and also to advance an agenda focused on preparedness and hazard mitigation. The work continues with support from the school's Parent-Teacher-Student Association. (Manning, 2012)

Another informal group, Parents for Preparedness (P4P), connects Portland-area parents to share school-level initiatives and to learn from local preparedness and seismic safety experts. P4P members are encouraged to explore roles in advocacy on school safety at the local school board and in the state legislature.

Oregon parents helped to supply the political support that led state legislators to approve $175 million in state funding for school earthquake retrofit projects in 2015-2017. Oregon's recent experience proves that individuals and informal groups can make a big difference in terms of awareness, school-level projects, and even the funding commitments needed to make schools safer.

6.3 Communicating with the Community

Communication is a fundamental component of preparedness, response, and recovery activities. It is thus important that schools conduct preparedness campaigns, special events, educational classes, social media dialogues, and other pre-disaster engagement before an event occurs to build a school- and community-based culture of safety.

Delivering a clear, consistent, actionable message through many trusted voices builds awareness and buy-in. Repetition drives the message home. Trusted messengers help to make the message immediate and relevant.

Some best practices for communicating with stakeholders include:

- Engage with both formal (designated, authorized) and informal (local trusted individuals) stakeholders.

- Use innovative and locally appropriate communication approaches to personalize outreach and make relevant to the wider school community.

- Link local community activities with ongoing or event-driven school disaster preparedness activities.

- Build sustained coordination and sharing of resources.

Key preparedness and response topics to communicate include:

- Notifications of planning efforts and opportunity for engagement;

- Awareness and educational messages about relevant hazards and risks;

- Explanations of emergency response protocols and procedures;

- Tools and technology platforms being used (or considered) by school officials and responders;

- School/community available resources in the event of a disaster; and

- Updates regarding ongoing mitigation or reconstruction projects.

6.3.1 Before the Event

As a school takes steps to understand, plan for, and reduce natural hazard risks, it is valuable to regularly communicate with a wide range of stakeholders. Communicating about hazards, risk, and emergency plans can be done as part of preparedness informational campaigns, while holding special events and classes, and through social media dialogues.

As school leaders, staff, students, parents, and others learn about the hazards they could experience and the potential consequences of hazard events, that knowledge should be shared with community members. It is also valuable to

> **Safety Information and the Rural-Urban Website Divide**
>
> A recent study of websites for every Colorado school district found a substantial rural-urban divide in terms of the information shared on the websites. Most rural schools had no emergency information available, at all, on their sites (Kaiser, 2016).

notify community stakeholders about planning efforts and opportunities for engagement and feedback. If the school is going to be used as an evacuation shelter, it is important to plan ahead of time with the community leaders on how the shelter will be operated and managed, how the school will be cleaned up, and how the expenses will be handled. The shelter planning helps expedite the resumption of school operations. Emergency response plans, protocols, and procedures should be shared publicly so that community members, especially parents, know what to expect and what they should do when a disaster happens.

If a school identifies a major safety concern, such as a serious structural deficiency in a school building, it is essential to communicate this information to the community in an honest, clear, and productive way. Many schools find that sharing information about the potential solutions (e.g., preliminary risk reduction options and associated costs) at the same time as information about the risks is actually reassuring to parents and other community members. However, the urgency of this information necessitates informing the community as soon as possible, rather than waiting for plans to address these safety concerns to be perfected.

Outreach and education to key school stakeholders ensures the wider community has up-to-date information on hazards, risks, and existing or developing emergency plans. Ongoing communications foster interactive exchange. Establishing these relationships builds a higher level of trust that the school is a trusted, meaningful, "go to" resource; this is an important identity if or when an incident occurs. Because many communities are multi-cultural, multi-lingual, and have a wide range of socio-economic diversity, having a communications strategy that utilizes a variety of distribution channels will have greater success. School communications will likely be integrated with the local law enforcement, emergency management, and other first responders. Coordinating with these and other stakeholders helps ensure the media receive the information that is needed. Partners can serve a vital role for supporting schools and releasing crisis information on their social media feeds and other technology platforms.

It is particularly important to have media policies and procedures set in place prior to any event. These should outline who should be the designated spokesperson on behalf of the school should an event occur, and set limits to media access to affected students and staff.

As demonstrated in prior chapters of this *Guide*, opportunities abound for school and community partners to participate in hazards preparedness, response, and recovery activities. Not only does the involvement of

6-12   **6: Engaging the Whole Community**   FEMA P-1000 community stakeholders improve institutional credibility on safety issues, it expands a school's ability to call on an expanded pool of resources and expertise when a disaster happens.

6.3.2 During the Emergency and Recovery Phase

Regular, honest, transparent communication with internal and external stakeholders is essential during the emergency, and throughout the recovery phase. Questions such as "How will this affect my child? When does school start again? Can you fix the school?" may be commonplace. Clear communication should inform the community about progress and successes, as well as needs, concerns and issues. Outreach reduces irresponsible and inaccurate speculation. It keeps stakeholders, particularly parents, focused on the learning programs and the school recovery process. In addition, having a communications plan can help streamline responses to numerous media inquiries that often follow.

The following are some tips for communicating after a disaster:

- Strive to provide timely and accurate information.

- Respond to questions thoroughly and honestly; do not be afraid to say "I don't know, but I will get back to you with an answer."

- Honor those who played a large part in the response and the recovery efforts, as well as those deceased, if any.

- Maintain ongoing, interactive dialogue between school and key stakeholders.

- Consistently emphasize a sense of caring and compassion.

- Quickly address rumors.

- Provide a primary point of contact and contact information for questions.

6.3.3 After the Recovery Phase

Following through with the community after the recovery phase is an excellent way to make a community more resilient to the next event. The opportunity to learn from the experience and share ideas on how to better prepare and respond next time is invaluable. The opportunity to reflect on how to do better and advocate for improvement is often received by an eager audience willing to share their ideas.

FEMA P-1000　　　　**6: Engaging the Whole Community**　　　　**6-13**

6.4 Tools and Technology for Effective Communication

School districts are expanding their use of tools and technologies for communicating alerts and warnings during the emergency phase, while also continuing to rely on these approaches during the longer-term post-disaster recovery period. Simultaneously, federal agencies and private sector companies are increasingly providing new solutions for timely and effective communication, particularly through the use of mobile devices.

Joplin Schools: A Case Study in Communications Success

On Sunday, May 22, 2011, an EF5 tornado hit the City of Joplin in southwest Missouri. The 200+ mph winds destroyed everything in the tornado's seven-mile-long path. The tornado caused 161 fatalities and approximately 1,371 injuries. The tornado struck the city of Joplin just after members of the Class of 2011 at Joplin High School had received their diplomas and were heading to post-commencement celebrations. While many key forward-leaning decisions factored into the extent of Joplin School's recovery, proactive communications strategies and innovative implementation tactics played a significant role.

Superintendent C.J. Huff, along with other district and volunteer leaders led the recovery and communication efforts in the months following the tornado. Building on pre-established communication tools, they set goals, created clear strategic messages and ensured consistent, timely responses to media. The Superintendent immediately formed a team of community professionals to meet daily, assess the district situation and priorities, develop solutions, communicate plans and act as a steering committee. School district leaders empowered key stakeholders to be associates for the schools. For example, teams of school communication professionals from the National School Public Relations Association (NSPRA) from outside of the area supported on-site operations during the early weeks after the disaster. Volunteer communications and community relations partners were recruited, engaged, and empowered to build and drive the district messages.

Communicators employed a wide range of outreach methods ranging from social media to the telephone. The media was provided unparalleled access. The City of Joplin used Facebook, Twitter, and YouTube to supplement information disseminated through traditional methods (FEMA, 2011). And when the media coverage began to taper off, the district created a new narrative—moving the message from the tornado's destruction to the power of recovery, featuring new schools being built and giveaway of free backpacks and school supplies. Hundreds of electronic press packets filled with message points and emotional human-interest stories and a three-day celebration script encouraged continued coverage. The result was that on the first day of the new school year—just 86 days after the tornado and right on schedule—every major cable news channel/network was there. Front-page stories in *USA Today* and *The New York Times* featured Joplin Schools (NSPRA, 2011).

Continuing to invest in communications, the district most recently transitioned to a new parent notification system using dialer messages, emails, and text messages.

Each school district is best positioned for identifying the method most effective for their target population; however, the most frequently used methods for communicating to the school community include:

- Loud speakers (internal to school);

- Newsletters sent to homes or distributed materials handed out or posted at the schools;

- Text messaging;

- Electronic automated telephone systems (e.g., informational hotlines, electronic phonetrees, robocalls, and interactive voice response systems);

- Emails;

- Websites;

- Social media (e.g., Twitter, Facebook, Instagram, Nextdoor, and Snapchat); and

- Mobile apps.

With the rise in use of mobile phones, particularly smart phones, social media is now a go-to information distribution resource. Social media platforms allow schools to deliver information rapidly, inexpensively, cover more geographic distance and more effectively target specific audiences than traditional media. During an event, it is essential to monitor, respond, and distribute information through social media platforms that are most prominently used in not only the local and regional community, but also nationally and internationally. School leaders should consider appointing a person or team to manage the school's social media engagement.

> **Rise of the Smartphone**
>
> "Nearly two-thirds of Americans are now smartphone owners, and for many these devices are the key entry point to the online world" (Smith, 2015). "Mobile devices accounted for 55% of Internet usage in the United States in January 2014." – CNN Money

---

Using Social Media to Support Disaster Recovery

One option for handling a school's social media channels is to pre-arrange a partnership with local social media-savvy Community Emergency Response Team (CERT) members or volunteers from the school's stakeholder community to virtually support the school's communications team in efforts to manage social media traffic.

This approach reflects one similar to that of a Virtual Operations Support Team (VOST), a team of trusted emergency managers and disaster volunteers trained in social media for emergency management. These individuals, located outside of an affected area, virtually support the operations of a local social media team operation. For more information, visit the Virtual Operations Support Group at http://vosg.us.

Regardless of the communication channel or response team approach, schools should track engagement (messages sent out, calls, press briefing) with all media.

In addition, in some communities, schools have hired communications outreach specialists to help manage and respond to media inquiries. In other cases, recovery leaders have held twice or once daily press conferences for weeks or even months after the disaster to ensure regular and systematic communication with the media, and hence with consumers of that media. Additional information on communications outreach is included in the *Resources Appendix*.

Chapter 7

Moving Forward

School leaders, staff, and teachers dedicate their lives to educating our next generation of leaders and workers. They also are tasked, on a day-to-day basis, with keeping children safe as they foster an environment conducive to learning. This is one of the greatest responsibilities assigned to adults in our society, and these efforts should be applauded and recognized.

This *Guide* has offered many practical suggestions for helping school leaders and other concerned parties to think about how they can best understand and ultimately reduce their risks related to natural hazards. The scope of this *Guide* was limited to guidance, tools, and funding mechanisms that are currently available, but it is clear that what is available is not enough to mitigate all school facilities in the nation in the near term. Accordingly, more research, funding, resources, and tools are needed to continue making our nation's schools safer.

The sections below identify some challenges and opportunities that were identified during the process of developing this *Guide*.

7.1 Identified Challenges

During the development of this report, the project team conducted a literature search of over 250 documents in an effort to share the latest available information and thinking with the reader. The process and findings from this study are documented in a companion document to this *Guide*; it is envisioned that the companion document will be published in late 2017. The literature search was conducted based on the three pillars defined in the Comprehensive School Safety Framework shown in Figure 7-1. Challenges were identified for each of the pillars: Safe Learning Facilities, School Disaster Management, and Risk Reduction and Resilience Education. The following same challenges were in common for all three topic areas:

- Budgetary constraints for building and maintaining disaster-resistant school facilities;

- Lack of coordination and cooperation between different stakeholder groups;

- Lack of stakeholder/decision-maker buy-in, involvement;

- Lack of access to expertise/experts;
- Lack of time; and
- Resistance to change or intervention.

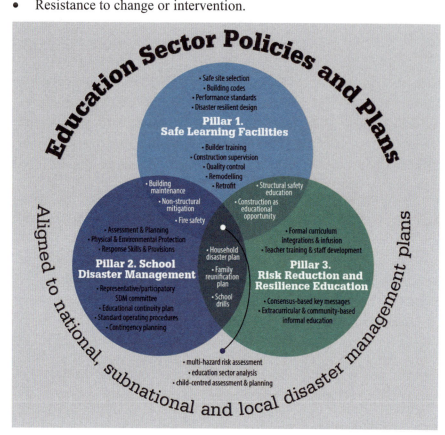

Figure 7-1 The three pillars of the Comprehensive School Safety Framework. This is an example of a topic-based approach to categorize the many activities that contribute to reducing school risk (GADRRRES, 2017).

Additional challenges that were raised during the development of the *Guide* were:

- Private schools tend to fall through the cracks. Many successful regulations and code requirements are only applicable to public schools and some charter schools. Approximately 5.2 million students attend close to 34,000 private schools throughout the United States (NCES, 2016). They make up over 10% of the student population in the United States and the safety of their school buildings from natural hazards should also be addressed.

- There is a lack of inclusion of disaster mitigation education in the common core curriculum in the United States. In particular, the literature review found that there were approximately 80% less resources in the United States when compared to international resources that address the Risk Reduction and Resilience Education pillar. As a comparison, 73% of the resources found under the Safe Learning Facilities pillar came from the United States. It was about evenly split for the School Disaster Management pillar. Because school curriculum is outside the context of this *Guide*, it was not addressed; however, parties that are interested and can have an impact over this topic should consider addressing this gap.

- Low community awareness of the possibilities of reducing risk through mitigation and planning is also an issue that was raised. As was mentioned throughout this *Guide*, many parents incorrectly assume that school buildings are safe from natural hazards. In cases where school leaders are aware of the potential risks they face, many feel powerless and that there is not much they can do to address these risks.

- Existing school buildings are often designated as emergency shelters without proper assessment of whether the buildings were adequately designed to resist natural hazard events and if they will be in adequate condition to shelter people during or after an emergency.

- Another big challenge is the changing U.S. education demographics and the implications of those changes for comprehensive safety strategies. Student populations are changing rapidly, and already more diverse (in every possible way) and lower-income than the U.S. population as a whole (Southern Education Foundation, 2015). This presents challenges for emergency plans and recovery (e.g., multiple languages, different family circumstances) and additionally, schools that have the riskiest school buildings and are in the riskiest areas are often the poorest schools with the least resources. Figure 7-2 illustrates the percentage of low income students in U.S. public schools per state.

7.2 Potential Opportunities

This *Guide* attempted to address some of these challenges described in Section 7.1 by providing guidance, resources, and examples illustrating how some communities have overcome similar challenges. Other potential opportunities to address these challenges are described as follows.

As the world becomes more connected, the disasters that strike one community in an isolated location are heard about in other parts of the world. Accordingly, there is a growing awareness of the impact natural hazards can have on schools and communities. Newly available tools and technology, such as social media and interactive online programs, are a strong driver of

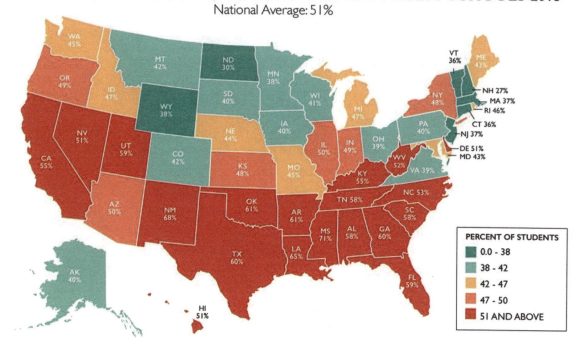

Figure 7-2   Map indicating the percentage of low income students in U.S. public schools (Southern Education Foundation, 2015).

this awareness. This advantage should be utilized to raise awareness among stakeholders and the broader school community. This is particularly useful for involving different partners in addressing school safety.

As science continues to improve, it also improves our understanding of our risks and ways to address them. Guidelines on the design and construction of tornado safe rooms and tsunami vertical evacuation structures are relatively new. More and more, building codes and regulations are requiring such refuges for new school buildings in communities at risk.

Finally, although natural hazard events can be devastating, they can also provide windows of opportunity when there is raised awareness and higher likelihood of political and financial support to address school natural hazard risks. Many significant improvements in building codes and school safety regulations and requirements have occurred soon after a disaster strikes. In many cases, these improvements affect areas that were not even impacted by the disaster.

Supplement E

Earthquakes

This supplement addresses existing and new schools located in earthquake zones in the United States and its territories. It provides guidance for existing buildings, including vulnerability assessment and mitigation (retrofit), and guidance for new schools that are in the planning stage. The guidance is based on field observations and research conducted on schools and other buildings that were affected by earthquakes worldwide.

> The United States Geological Survey (USGS) website (http://earthquake.usgs.gov/learn/) includes information on the causes and characteristics of earthquakes.

After reading this supplement, school administrators, school emergency managers, teachers, and other school leaders should be able to:

- Determine the level of seismicity in which their school is located;
- Understand key elements of seismic vulnerability assessments for existing schools and potential new school sites;
- Identify building design and mitigation options to achieve greater seismic resilience of school facilities, including potentially using the school as a postearthquake shelter;
- Create or update a school emergency operations plan addressing earthquakes; and
- Identify aspects that should be considered to facilitate school recovery following an earthquake.

E.1 Overview of Earthquakes

When the adjacent sides of pieces of the earth's crust (tectonic plates) suddenly slide with respect to one another along fault lines, an earthquake is created and waves are sent out in all directions. These waves cause the ground at any one site to rapidly move back and forth (shake) and this shaking is the primary cause of damage. The intensity of shaking at a given site is dependent mainly on the magnitude of the event and the distance of the site from the fault mechanism. There are many kinds of fault mechanisms created by various orientations of the edges of tectonic plates, as well as stresses in the plates away from the edges. These variations, along with differences in the surfaces along the slip, affect the characteristics of shaking, making every earthquake unique.

The ground shaking from earthquakes also causes landslides on unstable slopes and liquefaction (similar to quicksand) to occur in sandy soils that are saturated with water. Liquefaction under a building will cause large settlements and structural damage. If major motion occurs along a fault that is under water, great masses of water are set in motion causing large surface waves. When generated in the open ocean, these waves are known as a tsunami. When on lakes or partially enclosed bodies of water, they are known as seiches. See the *Tsunami Supplement* for information concerning these hazards.

E.1.1   Earthquake Impacts on Schools

The occupants of schools in the United States have been very lucky with respect to the risks from earthquakes. Since the great San Francisco earthquake of 1906, there have been 24 earthquakes in the United States that have been large enough, and near enough to population centers, to cause damage. Although building collapse and other life-threatening damage in schools occurred in many of these events, only three of these earthquakes occurred during school hours as shown in Table E-1, resulting in very few casualties. Although the records are inconsistent, two students have apparently been killed in earthquakes in the United States since such records have been kept, both in the 1949 Olympia, Washington earthquake from falling masonry (see Figure E-1).

> Gable walls—the triangular portion of walls under sloping roofs (which is seen missing at the top of Figure E-1)—are particularly susceptible to falling outward in earthquakes and are particularly dangerous if they are of heavy materials. In the case shown in Figure E-1, the gable was over an exit and the falling bricks (unreinforced masonry) killed a student.

Figure E-1   Site of a student fatality in the 1949 Olympia earthquake in Washington (NISEE, 1949).

Table E-1   Times of Damaging Earthquakes in the United States Since 1906

Year	Event	Magnitude	Time
1906	San Francisco, California	7.8	5:12 am, Wednesday, April 18
1925	Santa Barbara, California	6.8	6:42 am, Monday, June 29
1933	Long Beach, California	6.4	5:54 pm, Friday, March 10
1935	Helena, Montana	6.2	9:48 pm, Monday, October 18
1940	Imperial Valley, California	7.0	8:36 pm, Saturday, May 18
1949	Olympia, Washington	7.1	11:55 am, Wednesday, April 13
1952	Kern County, California	7.3	4:52 am, Monday, July 21
1963	Coalinga, California	6.4	4:42 pm, Monday, May 2
1964	Anchorage, Alaska	9.2	5:39 pm, Friday, March 27
1971	San Fernando, California	6.6	6:01 am, Tuesday, February 9
1980	Mammoth Lakes, California	6.2	9:33 am, Sunday, May 25
1983	Borah Peak, Idaho	6.9	6:06 am, Friday, October 28
1987	Whittier Narrows, California	5.9	7:42 am, Thursday, October 1
1989	Loma Prieta, California	6.9	5:04 pm, Tuesday, October 17
1992	Petrolia, California	7.2	11:06 am, Saturday, April 25
1992	Landers, California	7.3	4:57 am, Sunday, June 28
1994	Northridge, California	6.7	4:30 am, Monday, January 17
2001	Nisqually, Washington	6.8	10:54 am, Wednesday, February 28
2002	Denali, Alaska	7.9	1:12 pm, Sunday, November 3
2003	San Simeon, California	6.5	11:15 am, Sunday, December 22
2010	Eureka, California	5.7	4:27 pm, Saturday, January 9
2010	El Mayor Cucapa (Baja), California*	7.2	3:40 pm, Sunday, April 4
2011	Mineral, Virginia	5.8	1:51 pm, Tuesday, August 23
2014	South Napa, California	6.0	3:40 am, Monday, August 24

Note: Shaded rows represent earthquakes that occurred during school hours.

*While the epicenter of this earthquake was directly across the California border in Mexico, earthquake shaking impacted schools in Calexico, California.

Additionally, life-threatening damage and collapse of overhead items such as ceilings, light fixtures, and mechanical equipment has occurred in the majority of these events (see Figure E-2). These types of items are known as nonstructural components.

In addition to the risk of injury or death, earthquake damage often causes temporary closure of buildings and sometimes requires replacement. After the relatively modest Mineral, Virginia earthquake in August 2011, damage resulted in the closure of Thomas Jefferson Elementary and Louisa County High School. Although both were eventually replaced, the schools were not

> Nonstructural components refer to components in buildings that are not part of the structural frames and walls. These include architectural, mechanical, electrical, and plumbing systems, such as furniture, fixtures, equipment, and other contents.

Figure E-2 Examples of nonstructural damage. (Photo sources: (a) Baja Earthquake Reconnaissance Team, Earthquake Engineering Research Institute; (b) Gary McGavin, FEMA (2010a); (c) Russell Green, Geotechnical Extreme Events Reconnaissance Association)

available for several years. Repair or replacement of school buildings can take away a valuable community resource for long periods of time, and disrupt the lives of children and their families.

E.2 Is Your School in an Earthquake-Prone Region?

Risk from natural hazards is a combination of the severity of the hazard (e.g., location and intensity of an earthquake) and the vulnerability of the asset or institution under consideration (e.g., vulnerability of a school building). For earthquakes, the risk can be primarily mitigated by reducing the vulnerability (e.g., facilities designed or retrofitted for earthquakes) and also by thorough emergency planning. Effective mitigation depends on an understanding of vulnerabilities and their consequences.

> In 2017, FEMA ranked Puerto Rico, Alaska, California, Hawaii, Oregon, Washington, Utah, Nevada, South Carolina, and Tennessee as the top ten states/territories with the highest ratio of annualized earthquake loss to replacement value in their respective regions (FEMA, 2017b).

E.2.1 Determining the Severity of the Hazard

A relative seismic hazard map of the United States by counties is shown in Figure E-3. The map was developed in 2014 for FEMA P-154, *Rapid Visual Screening of Buildings for Potential Seismic Hazards: A Handbook* (FEMA, 2015b). It is based on work by the USGS to develop maps that are part of building code design procedures. These maps currently do not reflect recent increases in seismicity due to fracking or wastewater pumping ("induced earthquakes"). Current consensus of earthquake professionals is that such seismicity should not be included in seismic building codes because: (1) these earthquakes have been generally below the intensity level considered damaging by building codes; and (2) the seismicity will change location in accordance with pumping practices. Local jurisdictions should decide if building code provisions should change due to these local effects. Similarly, schools affected by seismicity considered temporary can also take precautions by including seismic issues in their disaster planning (see Section E.4) and mitigating obvious falling hazards (see Section E.3.3.3).

> There are several USGS papers on induced seismicity, including *2016 One-Year Seismic Hazard Forecast for the Central and Eastern United States from Induced and Natural Earthquakes* (Petersen et al., 2016), and *Myths and Facts on Wastewater Injection, Hydraulic Fracturing, Enhanced Oil Recovery, and Induced Seismicity* (Rubinstein and Mahani, 2015).

Locations designated to be in high or very high seismicity regions in Figure E-3 are at high risk of experiencing a damaging earthquake. Thorough plans should be developed by schools in these regions to understand potential consequences of earthquakes and to plan for them as a high priority. Areas in low regions of seismicity should set their priorities at preparing for other hazards before spending a lot of resources preparing for a damaging seismic event. The areas in moderate and moderately high regions of seismicity are only at moderate risk, but building types that have a high collapse risk (see Section E.3.1) should be identified for potential mitigation and obvious nonstructural risks, such a cabinets and bookcases (see Section E.3.3) should be anchored.

The USGS website includes an application that determines specific seismic parameters applicable for any given site defined either by coordinates or an address. These parameters can then be used to set the applicable region of seismicity for the site. The *Earthquake Appendix* provides instructions on the use of this application and interpretation of the results.

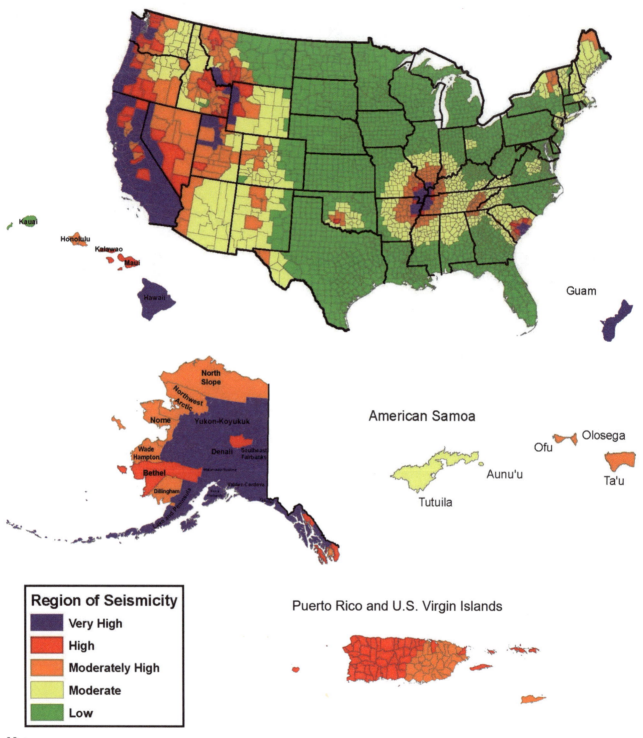

Notes:

(1) Based on NEHRP soil type B

(2) The seismicity at any site is calculated based on the highest seismicity at any point in a county. More accurate information on any site can be obtained from the USGS website: earthquake.usgs.gov/hazards.

(3) Islands not shown on American Samoa map (and their Region of Seismicity) are: (a) Rose Atoll (Low) and (b) Swains Island (Low).

Figure E-3   Relative seismic hazard map of the United States (FEMA, 2015b).

Earthquakes Do Not Just Affect the West Coast

The Mineral Virginia earthquake was a 5.8-magnitude event that struck on August 23, 2011, at 1:51 pm local time. The event epicenter was 38 miles northwest of Richmond and 5 miles south-southwest of the town of Mineral. Several aftershocks as large as 4.5-magnitude occurred in the following weeks, although the number of aftershocks was roughly a tenth of what one would expect in a West Coast event. This was the largest earthquake to strike east of the Rocky Mountains since 1944 and the largest earthquake in Virginia since 1897. It was felt from Alabama to Canada and as far west as Chicago; it was the most widely-felt earthquake in U.S. history. There were reports of significant damage to buildings near the epicenter and damage was documented in northern Virginia, southern Maryland, Washington, D.C., and as far away as New York City.

In Virginia, the Louisa County High School and an older elementary school were extensively damaged. Both buildings were older facilities that had been constructed prior to modern building codes. Classes had started that week and were in session when the earthquake struck. Fortunately, there were only minor injuries. However, damage to the two buildings was extensive and included significant cracking of unreinforced or marginally reinforced concrete block masonry walls. Nonstructural damage included collapsed ceilings and light fixtures and fallen shelves and cabinets, some of which blocked classroom exits. Damage was significant enough that officials quickly determined that the schools could not be reoccupied for the rest of the school year.

Next door, a middle school built in the early 2000s to a newer building code suffered only minor nonstructural damage and was able to resume classes the following week. Louisa County used that building in split shifts to hold many high school classes and the remainder were conducted in temporary modular classrooms, which were quickly brought onsite. The earthquake was eventually declared a federal disaster and the two damaged school buildings were demolished and replaced with new buildings using a combination of federal disaster funds and earthquake insurance, which the county had in place along with their normal coverage.

This event serves as a reminder that earthquakes can occur in many different parts of the U.S., even in areas designated as moderate or even low earthquake hazard. It is also a reminder that moderate earthquakes can cause significant damage to schools that are not designed for earthquake shaking.

E.2.2 Determining Your School's Vulnerability

Thorough emergency planning depends on an understanding of likely earthquake shaking levels, and expected damage to the site and facilities. Vulnerability depends on potential earthquake damage to: (1) school sites (from fault rupture, earthquake induced liquefaction, landslides, or poor access); (2) the school structure (structural walls, beams, columns, and foundations); and (3) nonstructural systems and contents of the school (for example, the nonstructural enclosure material on the building, interior partition walls, ceilings, lights, mechanical equipment, and bookcases).

Likely damage scenarios can best be determined by professionals after analysis of the site and facilities, but a general overview is given in the next section.

E.3   Making Buildings Safer

E.3.1   Existing School Buildings

In most cases, people interested in reducing the risk from natural hazards will be considering the safety of an existing school facility. Schools typically comprise more than one building, often built at different times. The earthquake vulnerability of school facilities should be assessed one building at a time. Potential vulnerabilities include issues with the site, issues with the structure of the building, which could lead to severe damage or even collapse, and issues with the nonstructural components and contents of the building, which can create falling hazards and internal disruption.

E.3.1.1   School Building Site

Although the vulnerabilities of the buildings are usually most critical, site issues should also be considered. Geotechnical investigations, done when the buildings were constructed, may be available, but may take professional interpretation. The local building or planning departments may also have maps that show local areas with a risk of fault rupture, landslide, or liquefaction at the site.

E.3.1.2   Vulnerable Building Types

Older Codes or Pre-Code Buildings. Buildings built in accordance with the current building code, including the seismic provisions, should provide adequate life safety from partial or complete collapse even in a very large earthquake (although that is not always assured, see Section E.3.2.2). However, the code has changed and improved over the last 50 years, so older buildings may not provide the same level of safety. In many parts of the United States, buildings built before 1990 may not have been designed with any seismic provisions. In general, buildings built in accordance with the 1976 *Uniform Building Code* (International Conference of Building Officials, 1976) or the *2000 International Building Code* (or later; ICC, 2000) should not have high vulnerability. Many school buildings have been built under very old building codes, or before any codes with seismic design requirements were in force. These buildings are typically more vulnerable than buildings built to modern building codes, and in some cases, pose a risk of collapse.

> *Geotechnical engineers* specialize in studying subsurface conditions at site, including conditions for building foundations and site hazards such as liquefaction and landslides.

> Life Safety is an engineering term used to describe a level of design. The main goal behind life safety is to prevent fatalities and serious injuries in a building due to failure or collapse of structural elements, such as columns and beams.

The facility manager of the school district or the local building department should be able to find what code was applicable when a particular school was built. To get a more refined evaluation of the probable earthquake performance of schools designed to various building codes, a structural engineer should be consulted. The *Earthquake Appendix* also has additional information about different building codes and their adequacy.

Vulnerabilities by Structural Type. Many older school buildings are built with a structural system called unreinforced masonry bearing walls, often abbreviated URM. This system uses brick walls, usually two or three bricks thick, with wood joists resting on the brick, forming the floors and roof. Some URMs use concrete floors instead of wood. The brick walls are heavy and brittle and not adequately tied to the floors and roof. URM buildings (that have not been seismically retrofitted) are almost always the first building type damaged in earthquakes. Chimneys, gable walls (see Figure E-1), and parapets (the low, barrier walls that extend at the edge of roofs) can collapse at low shaking levels, and with continued shaking, entire walls can collapse and sometimes the floors and roof. The severe damage in the 1933 Long Beach earthquake, which led to a strong statewide school safety law in California, was mostly in URM schools (see Figure E-4). As previously noted, the only fatalities to school occupants in the United States since 1925 were from the collapse of URM walls.

> "Schools shall be URM Free by 2033" — this policy statement released in 2016 by the Earthquake Engineering Research Institute highlights the needed actions to keep students safe from collapse-prone URM school buildings. More details can be found here: www.eeri.org /advocacy-and-public-policy /schools-shall-be-urm-free-by -2033/

It should be noted, however, that not all buildings with brick exteriors are URMs. For example, some more recent brick masonry buildings have steel reinforcing bars inside the wall. Other "brick-looking" buildings are only brick veneer, where one layer of brick (or even one-half of a brick) is attached to a backing wall of other material.

If it is suspected that a school is of URM construction, it is recommended that the exact construction type be confirmed by the facility manager, the local building department, or a structural engineer. For more information on URM construction, see FEMA P-774, *Unreinforced Masonry Buildings and Earthquakes* (FEMA, 2009c).

Another construction type that has proven to be susceptible to earthquake shaking is reinforced concrete built prior to seismic codes or according to very old codes (see Figure E-5). However, the vulnerability of this structural type is very dependent on the exact configuration and details of construction.

Precast concrete construction, for which pieces of the structure (e.g., walls, columns, beams) are cast separately, either on the site or in a factory, and then connected together to form the building structure, is particularly

Turning Disasters into Opportunities for Change

The 1933 Long Beach earthquake in southern California was a dramatic wake-up call. The magnitude-6.3 earthquake occurred at 5:54 p.m. on a Friday in March, just a couple of hours after children had been released from school. The scenes were shocking—hundreds of school buildings completely collapsed or were severely damaged, most which were of unreinforced masonry as shown in Figure E-4. Had the earthquake occurred a couple of hours earlier, thousands of students would have perished.

Fortunately, this close call resulted in action. One month after the earthquake, the California legislature passed the Field Act, which prohibited the construction of new unreinforced masonry buildings and essentially gave the State of California control over public school building design and construction inspection requirements. Since first enacted, the Field Act has had various amendments to improve its effectiveness and to address vulnerable existing public school buildings to date.

Figure E-4   Collapse of part of Jefferson High School in the 1933 Long Beach earthquake. (Photo source: Portland Cement Association)

The law has proven very effective in improving school earthquake safety in California (CSSC, 2009). However, state-of-the art of earthquake design has improved significantly since 1933 resulting in comparable changes to the building code and some building types built in certain periods of time are not currently deemed adequately safe. For example, some concrete buildings built before the mid-1970s are now considered dangerous, as discussed in Section E.3.1.2.

susceptible to damage. If it is suspected that a school is built of reinforced concrete and was built before 1980, a structural engineer should be consulted to more accurately identify the vulnerabilities.

Other construction materials, such as structural steel, reinforced masonry, or wood framing, are generally considered less vulnerable, but building performance in a real earthquake will be dependent on configuration and details.

Figure E-5 Collapse of a portion of a concrete school in the Helena, Montana earthquake of 1935 (NISEE, 1935).

E.3.1.3 Risk Reduction Measures

Risk reduction measures related to earthquake are often perceived as expensive and daunting projects. But the most effective and inexpensive initial risk reduction measure is to get a preliminary evaluation of the potential damage to the structural and nonstructural systems in a school so that appropriate planning for emergency earthquake response can be completed. Such a preliminary evaluation will also allow identification of priority and cost-effective measures to reduce the physical risks. Such a list of incremental improvements can result in steady risk reduction over time. This section covers risk reduction measures specific to the building structure. Section E.3.3 covers nonstructural risk reduction measures.

Seismic Evaluation. For emergency planning and for understanding and planning for physical risk reduction, it is imperative that the structural type and probable earthquake performance of school buildings be understood.

As previously discussed, by far the most dangerous structure type under earthquake shaking is the unreinforced masonry bearing wall building (URM). It is imperative that school districts, staff, and parents understand the risks from these buildings in moderate and high seismic zones and begin the process of risk reduction.

Other than determining the type of construction, the next level of seismic evaluation is a methodology developed by FEMA called Rapid Visual Screening, described in FEMA P-154. This method is intended to quickly

> In 2007, the State of Oregon screened nearly all K-12 schools and community colleges in the state according to FEMA P-154 (www.oregongeology.org/sub/projects/rvs/default.htm).

screen multiple buildings and rank them with respect to their relative risk. This method is intended to be used by personnel familiar with building construction and with special training in use of FEMA P-154. Such training is available from FEMA. These evaluations only take a few hours per building. The speed and efficiency is achieved at the expense of reliability, but the method is highly recommended as a first step in systematic evaluation.

Improvements in School Earthquake Safety: Examples from a City, a State, and a Province

The City of Berkeley, California recognized the potential vulnerability of older school buildings after the 1989 Loma Prieta earthquake and funded structural evaluations of several schools. Following a significant community effort, funds were raised, partly from the state and partly from local taxes, to either replace or seismically retrofit schools deemed as high risk.

Recently, the State of Oregon has adopted state policy on seismic safety in schools, including funding to start a risk mitigation program. In particular, state legislators approved $175 million in state funding for school earthquake retrofit projects in 2015-2017.

The province of British Columbia, Canada is also conducting a large program ($1.7 billion) to improve seismic safety of schools—all started by citizen action.

These highlight good examples of addressing school earthquake safety; however, many states and regions in high risk seismic zones have done little to either understand their risk or reduce it.

A more detailed evaluation can be done by structural engineers according to a national standard for seismic evaluation, ASCE/SEI 41, *Seismic Evaluation and Retrofit of Existing Buildings* (ASCE, 2014b). There are several different techniques for seismic evaluation in ASCE/SEI 41, with the level of effort ranging from three to four days to multiple weeks or even months. These methods will yield a much more detailed understanding of potential seismic performance, including identification of the specific structural deficiencies that could be strengthened.

Retrofit or Replacement. If seismic deficiencies are found in school buildings, particularly those that can be life threatening, studies should define at least preliminary fixes and approximate costs. Cost of seismic retrofits will vary a great deal due to differences in construction type and number and extent of deficiencies. When the estimated cost to retrofit reaches 50%-60% of replacement costs for the building, replacement will often be chosen. Besides the cost of retrofit, it must also be considered that the use of the building will be severely disrupted during construction and when completed, the building is still "old" in planning, appearance, energy efficiency, and access.

These issues may cause consideration of risk reduction by retrofit of only the most severe and dangerous deficiencies. For example, in URM buildings, parapets, chimneys, and gables could be braced to prevent their collapse and protection could be provided from falling objects at points of egress.

Partial or Incremental Retrofit. Due to the cost and disruption of complete structural seismic retrofit, partial retrofit is sometimes considered. Deficiencies to be mitigated using partial retrofit have to be carefully selected to be critical, or "first-to-fail." The concept is that for small to moderate earthquake shaking, the retrofit would prevent failure of the element and no other significant failures would occur. However, in stronger earthquake shaking, other significant failures could occur first, rendering the retrofit useless. For example, in URM buildings, parapets and appendages have been observed as the first to fail, and some communities have required parapets to be braced to achieve added street safety for small to moderate events. It has to be remembered, however, that for stronger earthquake shaking, failures from other deficiencies may occur; in the URM case, entire exterior walls may collapse and in some cases the floors and roof, as well.

Incremental seismic strengthening is similar to partial seismic strengthening except each increment is carefully planned to reduce risk and eventually lead to a complete retrofit. With careful planning, some increments can be defined and executed almost independently of others and be performed at time convenient to the occupants. For example, work at or around the roof can be performed at the time of normal roof replacement. Other work can be performed in rooms or areas of a building under remodeling or updating. The incremental concept requires detailed pre-planning and is most effective with stable facility management staffs, which is common in most schools. Incremental seismic strengthening of schools has been extensively described in FEMA 395, *Incremental Seismic Rehabilitation of School Buildings (K-12)* (FEMA, 2003).

Strengthening of one room or area of a school to serve as a "safe room" is not recommended because evacuation to that area during shaking is more dangerous than the "Drop, Cover, and Hold On" protocol described in Section E.4.1.1. Early warning systems that would provide adequate time for such evacuation are not fully developed and the early warning time will be too short for safe evacuation on most sites.

E.3.2 New School Buildings

E.3.2.1 Site Selection

All potential natural hazards should be identified and considered when selecting a site for a new school. Particularly for earthquake risk, sites on top of existing fault lines, on or below landslides, and on liquefiable soils should be avoided, if possible. Access and egress after an event should also be considered. The "softness" (loose soil or sand) or the "hardness" (firm soil or rock) of the site also has an effect on the earthquake shaking at that site, but these variations would seldom disqualify a site. Sites on a steep hillside may present landslide potential. A site study by an engineering professional specializing in soils and foundations (i.e., a geotechnical engineer) should be performed on any site being considered for a school. Sometimes, preliminary geotechnical reports become available as part of due-diligence studies for site purchase or selection. Often, alternate sites are not available, but once potential site hazards are identified, they can often be mitigated in the design and construction.

E.3.2.2 Building Codes and Expected Performance

Building Code Requirements. In the vast majority of communities in the United States today, new buildings are required to be designed and constructed according to the local buildings code, most often the *International Building Code* (ICC, 2014b). That code includes comprehensive provisions for earthquake shaking that specifies, among other things, the horizontal (lateral) strength of the building. Horizontal strength is important in earthquakes because the ground shaking induces horizontal loads on structures. For schools, three different levels of strength are described. For small occupancy (known as Risk Category II in the code), schools are required to be designed for the same earthquake loads as "normal" buildings. For more typical school buildings with an occupant load greater than 250 (known as Risk Category III), the required seismic loading is 25% higher than typical, intended to create a building with improved seismic performance, essentially, "safer." For schools designated as a shelter by the local community emergency authorities (known as Risk Category IV), the required seismic loading is 50% is higher than typical, intended to greatly increase the chances of the building being available as a shelter after an earthquake.

Expected Performance of Code-Compliant Buildings. The building code is basically intended to minimize the risk of death and serious injury, so Risk Category II buildings are expected to experience damage, but to protect life. In a major event, such buildings may not be useable or even economically repairable. In rare circumstances, an occupant or passerby might be injured by falling debris. For buildings in Risk Category III, being slightly stronger and better tied together, the risk to life safety is considered smaller, but the intent is not to assure the building is useable. For buildings designated as shelters, the obvious intent is to make them useable after most events, although there are not sufficient earthquake response data available to know if this can be guaranteed.

E.3.2.3 Schools as Shelters

Due to their locations throughout the community and availability of large spaces in gymnasiums and auditoriums, schools are often considered as ideal emergency shelters. However, for earthquake, most schools have not been constructed as a "designated" shelter according to the building code and therefore may not have adequate performance to serve as a shelter. When new school buildings are built, the district should coordinate with local emergency authorities to determine if the school is intended to serve as a shelter after an earthquake. If so, it should be classified as a Risk Category IV building, requiring stronger seismic design. However, shelter design goes far beyond a stronger structure, and may also require emergency generators, storage rooms for supplies, and other features desired by the local emergency services agency. Such provisions in schools will also significantly improve performance and minimize school closures due to damage (SEFT Consulting Group, 2015).

E.3.3 Nonstructural Systems and Contents

E.3.3.1 Building Code Provisions

Building code provisions to secure nonstructural items associated with the building itself, like partitions, ceilings, light fixtures, and mechanical equipment, have been in place for decades, particularly in high seismic zones. However, the physical installation of these items is normally controlled by the contractor and seismic protection features are often incorrect or incomplete. The contents of buildings, installed by the owner, such as furniture, storage shelves, and library stacks, are not controlled by the building code at all, and are typically just placed on the floor with no seismic anchors.

E.3.3.2 Vulnerabilities of Nonstructural Components

Considering the realities of installation as discussed above, it is no surprise that storage and bookshelves overturn and ceilings and light fixtures fall in schools in almost every earthquake. Many earthquake damage

Schools as Leaders in Community Resilience

Oregon faces a difficult challenge—it has a looming earthquake risk from the Cascadia Subduction Zone that was not identified as a risk until the mid-1980s. This means that most of the buildings, including schools, were not designed to properly resist the expected earthquake shaking.

In response to this, leaders and communities throughout Oregon have started to address their risk in various ways. The Beaverton School District is an excellent example of this. Following the 2014 approval of a major bond to help reduce school overcrowding and modernize schools, the Beaverton School District took this opportunity to design and construct seven new school buildings to a higher seismic standard than the code requires and that could also support their surrounding communities as emergency shelters. These leaders recognized that schools will have an important role in the response and recovery following an earthquake. As part of this effort, the Beaverton School District convened a workshop and subsequent meetings with various stakeholders to help inform these efforts. Figure E-6 illustrates a first floor plan of the high school. A report summarizing the resilience effort for the schools and community can be accessed here: www.beaverton.k12.or.us/depts/facilities/Documents/150710 _Beaverton%20School%20Report.pdf. (SEFT Consulting Group, 2015)

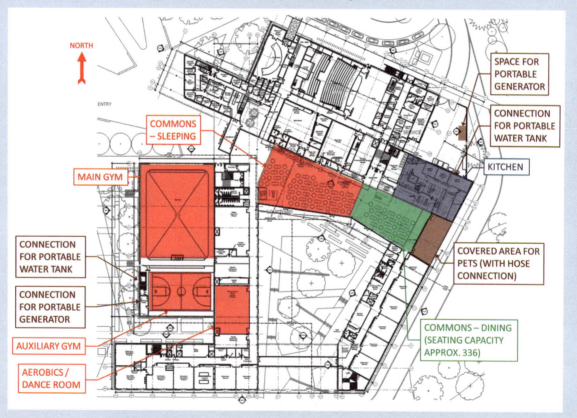

Figure E-6 High school first floor plan indicating the identified potential spaces for shelter sleeping and other important planning considerations for shelter operations (SEFT Consulting Group, 2015).

reconnaissance reports include observations that occupants of schools would likely have been injured if the event took place during school hours. Light fixtures are particularly prone to collapse and many such fixtures are sufficiently heavy to cause injury, as shown in Figure E-7. After the 1994 Northridge earthquake, the Los Angeles Unified School District had to replace nearly 4,000 light fixtures in 420 schools (personal communication, LAUSD). This vulnerability is a good reason for the "Drop, Cover, and Hold On" instruction that is the standard earthquake response taught to school children.

Figure E-7 Fallen light fixtures in a classroom after the Northridge earthquake. (Photo source: Earthquake Engineering Research Institute)

Unanchored shelving also invariably overturns, but this mostly contributes to a chaotic atmosphere and potential blocked exits rather than direct injury. It should be noted that shelving, cabinets, and other furniture brought into the school by the school staff are considered "contents" and these items (and their anchorage) are not covered by the building code.

Damage can also occur to overhead piping and mechanical ducting and fixed equipment such as boilers and air conditioners that will be costly and time-consuming to repair, possibly keeping the school closed for weeks. The costs may be covered by insurance or by certain FEMA programs, but documentation and records before and after the earthquake are critical as discussed in Section 5.1.2. It took twenty years for the Los Angeles Unified School District to "close the books" on damage from the Northridge earthquake even though no school was permanently closed by damage.

E.3.3.3 Nonstructural Risk Reduction Measures

Nonstructural seismic risk reduction measures can be separated into three categories: (1) very simple "do-it-yourself" activities; (2) handyman, or maintenance staff level activities; and (3) professional contractor level activities. Level 1 consists, for example, of moving furniture to clear egress routes or securing shelving to walls with small steel angles. Level 3, on the other hand, requires expertise in electrical or mechanical systems, roofing, or heavy construction (e.g., concrete work). A common exercise, both practical and educational, is to hold a "hazard hunt" in the classroom, particularly for Level 1 items.

> FEMA's Earthquake School Hazard Hunt Game and Poster is a great resource for school hazard hunts :
> https://www.fema.gov/media-library/assets/documents/90409

Many nonstructural elements that are susceptible to damage in schools, as well as potential retrofit measures for them are documented in the *Guide and Checklist for Nonstructural Earthquake Hazards in California Schools* (California Emergency Management Agency, 2011). Many of the items listed in that document are susceptible to damage, but are not necessarily life safety risks. It is important to understand the purpose of nonstructural retrofit and how they should be prioritized.

Bookshelves, Furniture, and Egress Routes. Portable cabinets and shelving should not be installed in main access corridors or where they could tip over and block doorways.

Wood or metal cabinets and shelving are often installed against a partition wall. Attaching them to the wall with light metal angles is simple and effective. Example details can be found in the *Guide and Checklist for Nonstructural Earthquake Hazards in California Schools*.

As previously noted, these items are generally not covered by the building code and it is the responsibility of school staff to anchor the items when initially installed or moved. These anchoring procedures should be put into school's operational plans.

Ceilings, Lighting, and Mechanical Diffusers. As previously mentioned, light fixtures have often fallen in classrooms during earthquakes. Light fixtures suspended on rods or chains can swing and be damaged or simply pull out of the ceiling mounting box. The dangers of such failures can be mitigated by installing safety chains or wires securely attached to the fixture and the structure above. Many classrooms have lightweight suspended ceilings made up of metal support bars and panels (called "lay-in" or "panelized" ceilings), with light fixtures also embedded within the metal bar system. Often the heating/cooling system runs above these ceilings and air vents are also embedded in the ceiling system. These systems have also proven to be very vulnerable to earthquake shaking if not installed properly with earthquake bracing. These systems often come apart when shaken, allowing relatively heavy light fixtures or mechanical vents to fall into the classroom space. Securing fixtures and vents and bracing these kinds of ceilings often requires the services of specialty contractors. Appropriate measures to secure these ceiling systems can be found in be found in the *Guide and Checklist for Nonstructural Earthquake Hazards in California Schools*.

> "Mitigation of Nonstructural Hazards in Schools" — this policy statement released in 2016 by the Earthquake Engineering Research Institute highlights the needed actions to keep students safe from nonstructural hazards in schools. More details can be found here: www.eeri.org /advocacy-and-public-policy /mitigation-of-nonstructural -hazards-in-schools/

Formal Evaluation and Systematic Retrofit. Other than the relatively obvious risks from shelving, cabinets, and overhead lights and ceilings, nonstructural seismic deficiencies that can result in costly damage and disruption must be identified with more comprehensive seismic evaluations. Although several documents include checklists that can be completed by staff (e.g., *Guide and Checklist for Nonstructural Earthquake Hazards in California Schools*), if occupancy immediately after an earthquake is desired—as for a designated shelter—a thorough evaluation, including proposed retrofits, is best prepared by an earthquake design professional. FEMA E-74, *Reducing the Risks of Nonstructural Earthquake Damage* (FEMA, 2012c) is a good resource for earthquake design professionals.

E.4 Planning the Response

The principles of developing school emergency management plans are well documented, particularly in the recently developed *Guide for Developing High-Quality School Emergency Operations Plans* (U.S. Department of Education, 2013). Earthquake-prone regions, in particular, require careful emergency planning as the range of impacts for a school or school district can be widespread. The size of the earthquake will affect both the severity of shaking (i.e., potential damage level) and the size of the shaken area (i.e., number of buildings or schools damaged). The vulnerability of school buildings will also have a major impact on emergency plans; planning for earthquake impacts on unreinforced masonry buildings may be very different than for modern buildings complying with earthquake codes. It is therefore highly recommended that at least preliminary evaluations of school buildings be performed and realistic scenarios developed that will identify likely impacts for planning purposes.

In addition to the general guidance on emergency operations planning provided in Chapter 4 of this *Guide*, unique conditions of the earthquake risk must be identified, such as:

- Can off-site earthquake damage to roads or structures diminish or eliminate site access?

- Is the site vulnerable to landslides or soil failure (e.g., liquefaction)?

- Are local emergency planners considering any of the schools as postearthquake shelter sites?

- Are there evacuation routes available from all areas of the buildings, free from potential overturned cabinets or shelving and safe from overhead falling hazards, particularly directly over exterior doors?

- Are postearthquake building safety inspections pre-arranged? Local structural engineers will be very busy and may not be immediately available after an earthquake without a prior arrangement. Are these inspections the responsibility of the individual school, the district, or the state? Are the buildings to be reoccupied without such inspections?

- Do emergency plans consider that the earthquake will likely not only damage the school facilities, but also much of the community?

E.4.1 During the Earthquake

Unlike many natural hazards, earthquakes strike without warning. Response must be immediate and automatic.

E.4.1.1 Recommended Protective Actions

Federal, state, and local emergency management experts all agree that "Drop, Cover, and Hold On" is the appropriate action to reduce injury and death during earthquake shaking in the United States. During earthquake drills, students and staff should practice the following steps:

- **Drop** down onto your hands and knees. This position protects you from falling down, but allows you to still move if necessary.

- **Cover** your head and neck with both arms clasping your neck with your hands. If a desk or table is nearby, crawl beneath it while keeping one arm over your head. If there is no shelter nearby, only then should get down near an interior wall (preferably in an inside corner) or next to low-lying furniture that will not fall on you.

- **Hold On** to your shelter while protecting your head and neck until the shaking stops. Be prepared to move with your shelter if the shaking shifts it around.

More details on the "Drop, Cover, and Hold On" recommendations, including for those with disabilities, and for situations other than a typical classroom, can be accessed on the Earthquake Country Alliance website (http://www.earthquakecountry.org/dropcoverholdon/).

In addition to regular school drills, the nationwide Great ShakeOut Earthquake Drills are an annual opportunity for schools, along with their communities, to practice protective actions during earthquake shaking. More information on the program can be found at www.shakeout.org.

E.4.1.2 Reactions to Avoid

During earthquake shaking, certain actions could be particularly dangerous. The following provide guidance on reactions to avoid.

- Do not run and stand in a doorway. Doorways are not particularly strong and doors could slam during shaking and cause injury.

- Do not run into a different room to find a shelter. It is difficult to run or move during shaking and may cause a fall. Moving around also exposes one to falling objects.

- Do not run outside. Exits often have heavy objects falling from above.

- Do not seek a "triangle of life." There is information circulating on the internet that students should be taught to lay *next* to a table or other solid object, rather than *under* it, as described by the "Drop, Cover, and Hold On" recommendation. Many organizations have rebutted this recommendation and the consensus is to teach the "Drop, Cover, and Hold On" response. (See also http://www.earthquakecountry.org /dropcoverholdon/.)

E.4.2 Immediately Following Shaking

Right after the shaking stops, an adult supervisor should check the egress path, both inside and out, for safety, and once safety is confirmed, students should be evacuated. Masonry falling hazards, such as parapets and gables (see Figure E-1), are often located over external doors and a visual inspection should assure that no obvious falling hazards are in the path of travel. Similarly, the outside assembly location should be checked for fallen power lines or other risks, including inclement weather conditions.

Seriously injured persons should not be moved unless there are signs of immediate danger, such as fire or the smell of gas. Instead, they should be covered with a sturdy table and send someone for medical help after the shaking stops. They should not be left alone, unless it is completely unsafe to have someone remain with the injured.

Once in the pre-selected assembly location, the pre-established post-disaster plan should be followed. The plan should consider the following:

> In general, the larger the mainshock, the larger and more numerous the aftershocks, and the longer they will continue.

- Aftershocks—earthquakes that follow the largest shock of an earthquake sequence—can occur for some time after the main earthquake. They can continue for a period of weeks, months, or even years. Students, teachers, and staff should stay away from buildings until they have been determined to be safe. This determination may have to be made by a professional engineer.

- Damage may have occurred over a large area, affecting emergency responders, parents, and the entire local school administrative structure. Damaged infrastructure and other conditions, such as roads blocked by rubble and debris, may impede immediate response and communication. It is critical to consider this in pre-arranged incident command structures and communications procedures.

- Access to the school may have been compromised. This condition should be included in contingency plans.

E.5 Planning the Recovery

After a strong earthquake, an engineer or building professional should assess the school facility, generally characterize damage or dangerous conditions, and determine if the facility is safe to re-enter and reoccupy. ATC-20-1, *Field Manual: Postearthquake Safety Evaluation of Buildings* (ATC, 2005), or similar documents and protocols will be useful. School districts may wish to have engineers under contract in advance so safety evaluations can be done quickly. It is important to note that safety evaluations are not the same as a property damage evaluation.

Once it is safe to access the school, there are two priorities: (1) documenting immediate postearthquake facility condition and damage (photos, videos, and field notes should be made); and (2) cleaning of the facility. Documentation will be needed for insurance and possible disaster assistance. Chapter 5 provides more detailed information on this topic. In some cases, schools will not be available for use for an extended period of time. This may be due to damage to access routes or to damage to the school building itself. This possibility should be included in contingency plans.

Most communities are unaware of the importance of schools in the community and their direct relationship to community disaster resilience. The school district should interact with community planners to maximize the effectiveness of joint recovery efforts. This is particularly true for earthquakes because they can damage and impact a large area, including much of the community, as well as many schools. The recovery and availability of schools is important to encourage return of the general population and re-establishment of normal activities. The importance of schools to community disaster resilience becomes particularly crucial when schools are intended to serve as emergency shelters following an earthquake. General advice and considerations for getting school back in session following a disaster is covered in Chapter 5.

E.6   Recommended Resources

Full citations of all references used to develop this supplement are listed in the References section in this *Guide*. The following is a list of recommended resources that might be useful for school leaders that are addressing school earthquake risk. In some cases, a document is both in the References section and listed here as a recommended resource.

Beaverton School District Resilience Planning. The report for the Beaverton School District from SEFT Consulting Group (2015) summarizes planning activities that were conducted to develop resilient design features design of two schools. These activities included the input of various stakeholder groups and design teams. https://www.beaverton.k12.or.us/depts/facilities /Documents/150710_Beaverton%20School%20Report.pdf

Case Studies of Seismic Nonstructural Retrofitting in School Facilities (Educational Facilities Research Center, 2005). Item by item list of nonstructural issues in school buildings with corresponding diagrams and instructions on retrofitting these items for improved safety. http://toolkit .ineesite.org/resources/ineecms/uploads/1055/Case_Studies_Seismic _Nonstructural_Retrofit.pdf

Classroom Education and Outreach Subcommittee. This subcommittee of the School Earthquake Safety Initiative (SESI) under the Earthquake Engineering Research Institute (EERI) has developed educational training

---

FEMA P-1000   **E: Earthquakes**   **E-23**

videos and design lessons for elementary schools with the purpose of promoting an ongoing dialog with parents, teachers, and administrators to develop advocates for earthquake school safety. To view available materials and videos, visit: https://www.eeri.org/projects/schools/subcommittees /#education.

Documentation for the 2014 Update of the United States National Seismic Hazard Maps (Petersen et al., 2014). Not specific to schools, but details the current state of hazard risk to areas of the country, identifying regions most at risk of earthquake. https://pubs.usgs.gov/of/2014/1091/

Earthquakes and Schools (National Clearinghouse for Educational Facilities, 2008a). A concise eight-page primer containing checklists that cover identification of vulnerabilities and appropriate emergency planning. http:// www.ncef.org/content/earthquakes-and-schools

FEMA P-154, *Rapid Visual Screening of Buildings for Potential Seismic Hazards: A Handbook* (FEMA, 2015b). This document presents a method of rapidly screening a group of buildings to determine their relative seismic risk. It is intended for use by personnel familiar with building construction, but not necessarily seismic experts. Training for its use is recommended and is available from FEMA. http://www.fema.gov/ media-library-data/1426210 695633-d9a280e72b32872161efab26 a602283b/FEMAP-154_508.pdf.

FEMA 159, *Tremor Troop: Earthquakes – A Teacher's Package for K-6. Revised Edition* (FEMA, 2000b). This teacher's package for grades K-6 provides ready-to-use, hands-on activities for students and teachers on the science of earthquakes and earthquake safety. http://www.fema.gov/media -library/assets/documents/2915

FEMA 240, *Earthquake Preparedness: What Every Child Care Provider Needs to Know* (FEMA, 2006a). Targets child care providers and features practical and low-cost techniques to make child care facilities safer in the event of an earthquake, whether they are based in a home or a larger facility. It offers tips for conducting earthquake drills and includes a checklist of supplies to keep on hand in an emergency kit.

FEMA 395, *Incremental Seismic Rehabilitation of School Buildings (K-12)* (FEMA, 2003). This publication was developed to provide school administrators with the information necessary to assess the seismic vulnerability of their buildings, and to implement a program of incremental seismic rehabilitation for those buildings. Different chapters are targeted at: (a) superintendents, board members and principals; (b) risk managers and financial managers; and (c) facility managers.

FEMA P-424, *Risk Management Series: Design Guide for Improving School Safety in Earthquakes, Floods, and High Winds* (FEMA, 2010a). This document includes extensive chapters on each hazard, explanations of structural and nonstructural seismic vulnerabilities found in schools and checklists covering site, structural and nonstructural issues.

FEMA 474, *Promoting Seismic Safety: Guidance for Advocates* (FEMA, 2005c). This document provides a collection of concise tips for advocates to utilize in gaining support for seismic safety policy changes and funding. https://www.fema.gov/media-library/assets/documents/3229

FEMA 527, *Earthquake Safety Activities: For Children and Teachers* (FEMA, 2005d). This document provides ready-to-use, hands-on activities for students and teachers explaining what happens during an earthquake, as well as how to prepare for and stay safe during and after an earthquake. http://www.shakeout.org/california/downloads/fema-527.pdf

FEMA 529, *When Earthquake Shaking Begins...Drop, Cover, and Hold On Poster*. This poster can be printed and used in schools. http://www.fema.gov /media-library/assets/documents/3266

Earthquake School Hazard Hunt Game and Poster (FEMA). This game and poster provides an interactive tool which engages young children to learn about earthquake hazard mitigation. https://www.fema.gov/media-library /assets/documents/90409

The Field Act and its Relative Effectiveness in Reducing Earthquake Damage in California Public Schools (CSSC, 2009). California passed the Field Act after the Long Beach earthquake in 1933 requiring state control of school design and construction, particularly related to seismic safety. This study clearly shows the effectiveness of this law over 75 years.

Guide and Checklist for Nonstructural Earthquake Hazards in California Schools (California Emergency Management Agency, 2011). This document includes a comprehensive checklist of nonstructural earthquake hazards typically found in schools and methods to mitigate the risks. Specific details are shown in sketches and references made to design professionals when needed. http://www.caloes.ca.gov/PlanningPreparednessSite/Documents /Nonstructural_EQ_Hazards_For_Schools_July2011.pdf

Great ShakeOut Earthquake Drills. These drills are an annual opportunity for people in homes, schools, and organizations to practice what to do during earthquakes, and to improve preparedness. These drills are practiced in many areas of the United States and around the world. The shakeout site includes many resources for school earthquake drills and education, including templates, checklists, activities, etc. teaching materials. www.shakeout.org

Incorporated Research Institutions for Seismology. This site has educational resources, including interactive maps of recent seismic activity that can be used as a teaching tool in classrooms. https://www.iris.edu/hq/

Risk RED School Disaster Response Drill Model and Templates. This site provides curriculum guidance to be used in schools for disaster preparedness. Templates for drills, including letters to parents, step by step activities, forms and ICS procedures, forms and surveys. www.RiskRED.org/schools.html

School Disaster Readiness: Lessons from the first Great Southern California Shakeout (Risk RED, 2009). Explains process of utilizing the shakeout curriculum and the success. Provides step by step information and activities, checklists, and case study information. http://www.preventionweb.net/files /14873_RR2008SchoolReadinessReport.pdf

School Facilities Improvement Program. Northern California Chapter of the Earthquake Engineering Research Institute. This page provides examples of seismic mitigation programs related to educational facilities. http://www .eerinc.org/?page_id=240

USGS Website. This website provides many resources related to earthquake hazards. https://www.usgs.gov/

E-26 E: Earthquakes FEMA P-1000

Supplement F

Floods

This supplement addresses existing and new schools located in flood-prone areas of the United States and its territories. It provides guidance on evaluating flood vulnerability and mitigation options for existing school facilities, and site selection and design guidance for new schools that are in the planning stage. The guidance is based on field observations and research conducted on schools (and other buildings) that were affected by floods. The corresponding *Flood Maps Appendix* provides additional, more detailed information on understanding and using flood hazard maps. Much of the information in this supplement and the *Flood Maps Appendix* also pertains to schools that are prone to hurricane storm surge flooding.

> "Flood-prone" means a normally dry area that can be inundated by water during an inland or coastal flood. "Flood-prone" includes "storm surge-prone."

After reading this supplement, school administrators, school emergency managers, teachers, and other school leaders should be able to:

- Know where to find flood hazard maps, and determine if their school facility is in a flood-prone area;

- Understand key elements of flood vulnerability assessments for existing schools and potential new school sites;

- Understand how to improve flood resistance of existing schools, and recognize the importance of incorporating flood-resistant design in new facilities to achieve greater resilience;

- Create or update a school disaster plan with specific considerations for floods; and

- Identify aspects that should be considered to facilitate school recovery following a flood.

F.1 Overview of Floods

Flooding is a condition where water moves beyond normally wet areas and temporarily inundates or saturates normally dry areas. School facilities have been and will continue to be affected by flooding. Figure F-1 provides an example of a school impacted by a flood. Effects can span from minor damage and short-term disruption (hours to days), to major damage and need for reconstruction or relocation (months to years).

Figure F-1   Ramstad Middle School, in Minot, North Dakota was flooded in June 2011, badly damaged, and ultimately decommissioned in June 2012. A replacement school was constructed in a less floodprone area.

> Every community — coastal, piedmont, mountain, plains, or desert — has some areas subject to some type of flooding.

Floods are the most common natural hazard and occur in every state and territory in the United States. Floods can result from runoff from excess rainfall or snowmelt; ice or debris blockage of streams and drainage; high tides and storm surges (see the *Hurricane Supplement*); tsunamis (see the *Tsunami Supplement*); or failure of levees, flood protection structures or dams.

Floods can occur with days of warning or without warning. Those with days of warning allow time to implement plans to ensure safety of school occupants and to reduce flood damage. It is more difficult to respond to those with little or no warning, which require planning in advance through facility design and emergency operation plans.

F.1.1   Flood Impacts on Schools

> In August 2016, numerous schools in Louisiana were closed due to severe flooding. In one school district, 17 schools were significantly damaged while 6 of those buildings were completely flooded (Mitchell, 2016).

Flood effects on schools and communities can be severe, particularly for children who experience flooding. Studies have documented the effects of flooding on schools, including displacement and school interruption for students, and highlight the importance of having both functional schools and schools that are supportive of students following flood events (Walker et al., 2010; Fothergill and Peek, 2015).

In particular, *Children and Young People 'After the Rain Has Gone' – Learning Lessons for Flood Recovery and Resilience* (Walker et al., 2010) documented how the recovery process, with daily disruptions, affected children following the heavy rainfall and severe flooding that affected Kingston-upon-Hull, United Kingdom, where 91 of 99 schools closed for some period of time. The report provides recommendations on how schools and the educational system can better support students during long-term flood recovery.

F.2   Is Your School in a Flood-Prone Region?

Knowing whether a school site is susceptible to flooding is important for both existing schools and for siting new schools. Knowing the vulnerability of existing schools to flooding is important for developing mitigation strategies and emergency plans. Knowing the vulnerability of potential school sites is important when different sites are compared and when school facilities are designed.

> Just because a site has not flooded in the past does not mean that you can assume it to be safe from flooding in the future.

When evaluating flood vulnerability of a particular site, all of the following should be considered:

1. **Review of flood hazard maps.** Do flood hazard maps indicate that the school site or facility is within or adjacent to an area subject to flooding? What information does the map (and any supporting study) provide and what was the intended use of the map?

2. **Review flood protection structures.** Is the site located behind a levee or seawall or does the site safety rely on other flood control structures? Has that levee or structure been accredited as providing a certain level of flood protection? What entity is responsible for inspection, maintenance and repairs? Flood hazard maps provide authoritative information from government agencies, but mapping consideration for levee protection is not always uniform.

> The 100 counties with the highest composite flood risk scores—as defined in a study by The Pew Charitable Trusts—are in 23 states and have 6,444 schools serving almost 4 million students. Louisiana has the highest number of these counties, with 24 in the top 100 (Pew, 2017).

3. **Identify past flooding events.** Has the school site or facility flooded before? Has the surrounding area flooded? When, how often, and what were the circumstances? Experience is important because flood hazard maps may not show local drainage issues or flooding from small watersheds.

4. **Identify relevant trends and development.** Are there trends showing that occurrence and/or severity of floods are increasing? Is future land development likely to increase runoff and flooding? Is sea level rising in the community? Is the ground subsiding? Is there an increase in "nuisance flooding" (see Figure F-2)? Answering these questions can help determine if past flood trends will continue, or if flooding is likely to worsen.

FEMA P-1000   **F: Floods**   **F-3**

Nuisance Flood Events on the Rise

Sea levels are rising and flood hazards are increasing in many areas. A recent technical report found that all three coasts in the United States have seen a 300 to 925% increase in nuisance flooding since the 1960s (NOAA, 2014).

Figure F-2　Graphic illustrating the rise in nuisance flooding around the United States, but especially off the East Coast (NOAA, 2014).

F.2.1　Flood Hazard Maps

This section provides an overview of flood hazard maps and how to use them. For detailed information on understanding and using flood hazard maps, see the *Flood Maps Appendix*.

> A Flood Insurance Study (FIS) Report is produced by FEMA along with the FIRM. The FIS contains a summary of the modeling and mapping information used to create the FIRM, and will contain additional detailed information that is needed for flood vulnerability assessments and flood-resistant design.

The most common flood hazard map is the Flood Insurance Rate Map (FIRM) produced by FEMA. FIRMs have been produced for over 21,000 communities in the United States and its territories, and should be the first stop when flood hazards are investigated. Most states and communities regulate development using flood hazard zones and flood elevations shown on the FIRM. This is done through floodplain management regulations, and through building codes. Regulations and building codes both will reference the flood hazard map adopted by a community, which most often is the FIRM.

FIRMs can be obtained from a number of sources. The local government is a good place to check. Many communities and states display FIRMs and related flood information on their websites. In addition, FIRMs can be obtained from: (1) FEMA's Map Service Center, an online repository of flood hazard map and flood study information (https://msc.fema.gov/portal);

and (2) FEMA's National Flood Hazard Layer, an online digital flood map of the nation (https://www.fema.gov/national-flood-hazard-layer-nfhl).

FIRMs will indicate important information, such as flood hazard zones (see *Flood Maps Appendix* for details) and Base Flood Elevations (BFEs). BFEs indicate the expected elevation of the flood surface during the base flood, also known as the 1% annual chance flood or the 100-year flood.

> The term "100-year flood," also known as the base flood, is misleading. It does not mean a given flood elevation will be reached only once every 100 years. It means there is a 1-in-100 chance (1%) of a flood reaching that elevation in any given year.

It is important to be aware of the following mapping assumptions and limitations in developing FIRMs:

- FIRMs depict flood hazards at the time the flood study was performed, and if significant time has elapsed, current land and flood characteristics may have changed.

- FIRMs show flood extents and elevations associated with floods with a 1% chance of occurring in any year (the "100-year" flood). More severe floods can and do occur.

- FIRMs do not account for flood-related hazards like levee or dam failure, mudslides, debris flows, channel migration, and shoreline erosion. These hazards can be destructive, as shown in Figure F-3.

- FIRMs do not account for future changes in development, hydraulics and hydrology, sea level rise, climate, and other factors. A complete evaluation of existing school facilities and plans for new school facilities should take these factors into consideration, which will require information that is not provided by a FIRM.

> FIRMs and FISs do not tell you everything you need to know about flood hazards. Other flood hazard maps (e.g., showing storm surge limits, areas subject to sea level rise, dam inundation areas, areas of future riverine flood hazard) should be reviewed. See *Flood Maps Appendix*.

Figure F-3   This school in Puerto Rico was undermined by riverine erosion during Hurricane Georges in September 1998 (FEMA, 1999).

Flood hazard zones and BFEs are important because:

- New school facilities must be designed to meet special requirements that vary with flood hazard zone.

- The lowest floor of a new school facility must be elevated or floodproofed to at least the BFE.

- Zones and BFEs help guide evaluation and modification of existing schools to achieve greater flood resilience.

F.2.2 Levees and Other Flood Control Structures

Levees and flood control structures have been constructed in many areas of the United States, but not all provide the same level of protection, and all will be subject to failure under certain conditions. If a school site relies on these types of structures for flood protection, information about the protection structure should be reviewed before committing to school construction or major repairs.

The book *So, You Live Behind a Levee!* provides more information on levees and their associated risk (ASCE, 2009). FEMA's Levee Analysis and Mapping Procedure (LAMP) website (http://www.fema.gov/final-levee-analysis-and-mapping-approach) provides the latest information on how levees are incorporated into FIRMs. Levee information available from state water resources agencies, levee districts, and the United States Army Corps of Engineers should be reviewed.

F.3 Making Buildings Safer

The single biggest determinant of flood vulnerability is elevation—both ground elevation relative to flood elevation, and building floor/basement elevation relative to flood elevation are important. Knowing and using this information is crucial to having strong and safer school facilities. In the context of flood, "safer buildings" means:

- Wherever possible, school buildings and associated facilities—including access roads and equipment storage facilities (see Figure F-4)—should remain dry during design flood conditions (i.e., ground elevations should be higher than the design flood elevation).

- If any portion of a school site gets wet during flooding, school occupants should not be in danger and any portion of a wet facility should sustain no more than minimal flood damage. Minimal flood damage means only clean-up and cosmetic repairs would be required and time out of service would be short.

> "Design flood" is the technical term used for the flood level used for design purposes, as dictated by the building code or state/community floodplain regulations. The design flood will always be equal to or greater than the base flood.

Figure F-4 School buses in New Orleans, Louisiana were swamped by the floodwaters following Hurricane Katrina in September 2005 (Photo source: Liz Roll, FEMA).

- Downtime and damage should be minor, even if a flood exceeds the design flood by a small amount. If a flood greatly exceeds the design flood, major damage should be expected and plans for using alternate school facilities should be considered.

- Facilities should be useable immediately after a design flood.

- If the school is designated as a shelter for flood, the shelter portion of the school should be operational during the design flood.

Many existing school facilities do not meet the criteria listed above, and it may be much easier to design new schools to meet the criteria than to modify existing schools to meet them.

The best time to take steps to reduce flood damage and manage any potential impacts to facilities and operations is long before a flood. Flood defense schemes can be complex, require significant manpower and material, and take time and planning to implement. Principal ways to eliminate or reduce flood damage include:

- **Avoidance.** This involves locating or relocating a school facility so it is on high ground, outside the area that is prone to flooding.

- **Elevation.** This involves constructing a new school facility or retrofitting an existing facility so the lowest floor and all important spaces are above the design flood elevation.

- **Flood Protection.** This involves constructing flood protection structures (e.g., levees, berms, or floodwalls) or modifying the site around the school building (e.g., improving drainage or constructing diversion channels) to reduce the likelihood of flood damage. It is important to note that if the flood level exceeds the protection level afforded by these measures, flood damage could be major.

- **Floodproofing.** This involves any combination of structural or nonstructural adjustments, changes or actions to a school facility within the flood hazard area that reduces or eliminates flood damage to the structure, its contents, and its attendant utilities and equipment. Floodproofing can be divided as follows:

  o **Dry Floodproofing.** This involves making the structure (including attendant utilities and equipment) watertight, with elements substantially impermeable to the passage of water, and with structural elements capable of resisting flood loads. Dry floodproofing is permitted for non-residential structures only.

  o **Wet Floodproofing.** This involves allowing portions of a structure to intentionally flood, but constructing the structure with flood-damage resistant materials.

  o **Active Floodproofing.** This involves temporary measures that require human intervention to carry out, install, or deploy. Examples include: temporary relocation or elevation of valuable records (e.g., not placed in basements), computer equipment and other items; use of sand bags or other temporary flood barriers; and closing of flood doors and gates.

  o **Passive Floodproofing.** This involves permanent measures (e.g., waterproof coatings, or floodwalls) that are built-in and require no human intervention to deploy in the event of a flood. Passive floodproofing is generally more reliable than active floodproofing, and requires no flood warning time and human action to deploy. However, passive dry floodproofing installations may affect or interfere with function and operations.

> Schools may be floodproofed using dry and/or wet methods, but local jurisdictions should be contacted for exact requirements.

> Active and passive floodproofing measures may be permitted for existing and new schools, but local jurisdictions should be contacted for exact requirements.

F.3.1 Existing School Buildings

Existing schools will offer a wide variation in flood resistance, depending on age, location, and original design. Some existing schools may meet the criteria listed in Section F.3, but many will not. In some cases, flood vulnerability may be unknown or known only qualitatively.

School leaders should investigate the following questions about flood vulnerability of an existing school:

- Is the school wholly or partly within a mapped flood hazard area?

- Are specific flood hazards and potential flood damages known for an existing school?

- Is it cost-effective to modify or retrofit an existing school to address flood vulnerability? Should wet floodproofing, dry floodproofing, or a combination of the two be used?

- What level of flood protection is justified based on site and budget constraints? When should the work be done, before or after a flood?

- Will FEMA and community regulations for substantial damage and substantial improvement be triggered (i.e., requirements that existing buildings be made compliant with current flood hazard map and building code provisions)? Local floodplain managers or building officials should be consulted about substantial damage and substantial improvement issues. Section 5.1.3 provides more detailed information on substantial damage and substantial improvement triggers.

- Will outside funding stipulations require compliance with the current flood hazard map and building code (even if damage and/or work do not constitute substantial damage or substantial improvement)?

- Are there operational considerations that would affect decisions regarding flood retrofits and flood protection? Section F.4 provides flood-specific considerations for emergency operations planning.

- Should existing schools be replaced (if so, where, when, how)?

In some instances, the decision will be made to decommission and replace a flooded school, as in the school shown in Figure F-1. In other cases, the decision will be made to repair and retrofit a flood-damaged school, as shown in Figures F-6, F-7, and F-8).

Acquiring and evaluating the information listed above will likely require involvement of designers and contractors with specialized flood and/or educational facility experience, and will require considerable effort. However, modifying existing schools to achieve increased flood resilience may also provide opportunities to expand or reconfigure the school, and to modernize and improve function and operation. The evaluation of an existing school will require information, as follows:

FEMA P-1000   **F: Floods**   **F-9**

Danville Middle School: Retrofit to Reduce Flood Damage

The remnants of Tropical Storm Lee dumped approximately one foot of rain over parts of eastern Pennsylvania in September 2011. The Danville Middle School sustained flooding from a few inches to approximately six feet deep, as shown in Figure F-5. Since the school was not "substantially damaged" by the flooding, the school district looked at the options—that is, whether to repair and wet floodproof the school (design to allow flood water to enter) so it would avoid structural damage and could be cleaned easily, or to dry floodproof (keep flood water out of) the school. (It is important to note that had the school been "substantially damaged" as defined by the code, wet floodproofing would not have been permitted by the National Flood Insurance Program and building code.) Dry floodproofing the entire school was not financially feasible so a combination of measures was undertaken, as follows:

- Some portions of the facility were protected by walls or removable flood gates at entry doors.
- Other portions of the facility were wet floodproofed.
- Heating, Ventilation, and Air Conditioning (HVAC) equipment was elevated.
- The lower portions of walls were removed and replaced with flood-damage resistant walls, as shown in Figure F-6.
- Carpet and vinyl floors were replaced with sealed epoxy flooring.
- Metal lockers were replaced with high density polyethylene lockers, as shown in Figure F-7.
- Over 200 flood vents were installed to allow water into the building, as shown in Figure F-8, relieving potential pressure from floodwater and structural failure that complete dry floodproofing might cause.

The clean-up and restoration work at the Danville Middle School took approximately two years and cost approximately $12 million.

Figure F-5   Danville Middle School in Pennsylvania flooded in September 2011. (Photo source: Reynolds Restoration Services, Inc.)

Danville Middle School: Flood Retrofit Features

Figure F-6 Wall repair and reconstruction to resist future flood damage. (Photo source: Reynolds Restoration Services, Inc.)

Figure F-7 Metal lockers were replaced with high density polyethylene (HDPE) lockers (Scranton Products, 2016).

Figure F-8 Flood vents were installed throughout most of the school. Allowing floodwater to enter prevents structural failure of walls if water is outside but not inside (Smart Vent, 2016).

- Construction plans for the original construction, and information regarding any major repairs or modifications since original construction should be reviewed. Structural drawings, architectural drawings, and information on mechanical-electrical-plumbing systems will be especially important.

- Top of floor elevations (in buildings and basements) that are potentially susceptible to flooding should be identified. Lowest elevations of utility equipment will be needed and locations of ductwork, chases, piping, and wiring will be useful.

- Locations and elevations of any possible points of floodwater or groundwater entry into the facility (e.g., doors, windows, above- and below-ground connections to other buildings, air vents, floor drains, and wall or slab penetrations for utilities) should be identified.

- Documentation for any prior flooding at the site or in the school facility should be investigated, including the following for each known flood event: source of flooding; flood elevation and duration; occurrence of other flood-related hazards (e.g., waves, erosion, mud, and debris); type and extent of flood damage sustained; and cost of repairs.

- The FIRM, FIS, and other available flood hazard maps and information should be reviewed.

- State and local floodplain management and building code requirements that could affect any work on the facility should be identified.

F.3.2   New School Buildings

New schools must be planned, sited, designed, and constructed in accordance with all applicable building codes, flood regulations, and land use plans. In many cases this will mean in accordance with the state building code and local flood regulations. In some cases, school construction may be governed by a state school construction code, which may differ from the building code.

Wherever possible, designing and constructing new facilities beyond areas shown as inundated on the flood hazard map should be considered. If located inside a flood hazard area, designing and constructing for flood levels above the minimum requirements set by regulations and building codes, which may have building elevation and other design requirements that exceed minimum National Flood Insurance Program requirements, should be considered.

F.3.2.1   Building Codes and Standards

In most cases, state building codes or school construction codes will be based on a "model" building code, such as the *International Building Code* (ICC, 2014b), and certain design standards referenced by the model code, such as ASCE/SEI 7, *Minimum Design Loads for Buildings and Other Structures* (ASCE, 2010) and ASCE/SEI 24, *Flood Resistant Design and Construction* (ASCE, 2014a). These codes and standards typically classify buildings by their importance, or by the risk to occupants in a design-level hazard event.

F-12   F: Floods   FEMA P-1000

Engineers use ASCE/SEI 7 to determine flood loads that buildings should be able to withstand. Typical school buildings with a capacity of more than 250 people (known as Risk Category III in ASCE/SEI 7) must be designed to withstand greater flood loads than most residential and commercial structures (known as Risk Category II in ASCE/SEI 7). This is intended to result in school buildings that are more resistant to flood loads and are essentially "safer." Schools designated as an emergency shelter by the local community emergency authorities (known as Risk Category IV in ASCE/SEI 7) must be designed to withstand even greater flood loads, with the intention of increasing the chances that the building will be available as a shelter after a flood.

> Chapter 5 of FEMA P-424, *Risk Management Series: Design Guide for Improving School Safety in Earthquakes, Floods and High Winds* (FEMA, 2010a), contains detailed information about flood-resistant design of new schools.

Engineers use ASCE/SEI 24 to determine elevation and floodproofing requirements for schools inside mapped flood hazard areas. All schools, regardless of capacity, are designated as at least Flood Design Class 3, which requires schools to be elevated or floodproofed to levels above those required for Flood Design Class 2 buildings (most residential and commercial buildings). Elevation or floodproofing above the BFE is known as adding "freeboard." Freeboard is required for school buildings not used as emergency shelters (Flood Design Class 3 buildings) and additional freeboard is required for school buildings used as emergency shelters (Flood Design Class 4 buildings).

In summary, these designations are important for two reasons:

- School buildings (designated as Risk Category III or IV structures) must be designed to withstand greater flood loads than most residential and commercial (Risk Category II) structures.

- Schools inside mapped flood hazard areas have more stringent elevation and floodproofing requirements than most residential and commercial buildings.

F.3.2.2 Federal Executive Orders Related to Flood

Some schools may be subject to federal Executive Order (EO) 11988, which mandates certain "critical" facilities be located outside the area subject to flooding during the 0.2% annual chance (500-year) flood. EO 11988 does contain some exceptions to the siting requirement, and may require elevation or floodproofing to the 0.2% annual chance flood elevation instead. More information on application of EO 11988 can be found in Section 5.1.6.3 of FEMA P-424, *Risk Management Series: Design Guide for Improving School Safety in Earthquakes, Floods, and High Winds* (FEMA, 2010a), and in

Section 2.1.4 of FEMA P-936, *Floodproofing Non-Residential Buildings* (FEMA, 2013a).

In 2015, Executive Order 13690 was issued, revising EO 11988 and establishing a new Federal Flood Risk Management Standard (FFRMS) http://www.fema.gov/federal-flood-risk-management-standard-ffrms. FFRMS requires federal agencies to consider current and *future* risk when taxpayer dollars are used to build or rebuild in floodplains. Applicability of FFRMS to a new school must be determined on an individual basis through consultation with federal agencies involved in school construction and funding.

F.4   Planning the Response

Flood-specific issues related to developing emergency operations plans are discussed in this section. For more general aspects of emergency operations plans, see Chapter 4.

It is recommended that administrators and emergency managers consult with people who have been through past floods and learn from their experience. What did their experience teach them? What would they do again, or do differently? This could be one of the most important and revealing pre-flood actions a school could take to help inform the development of effective emergency operations plans.

Once a flood watch or warning is issued, or once a flood is known to be imminent, the focus will be on: communicating with parents and the community; safe release of students and staff; orderly shut-down of operations; implementing any active floodproofing measures; evacuating or elevating contents, equipment and moveable, high-value objects; disconnecting utilities; and securing the facilities. All of these considerations should be addressed in school emergency operations plans.

Generally, school facilities in flood-prone areas should be evacuated prior to a flood. Even schools protected by local berms, levees, and floodwalls should evacuate during flood conditions. There may be exceptions, but people should occupy schools during floods only when there are no other feasible options. Exceptions might include: (1) where certain facilities have been designated as hurricane or tsunami refuges of last resort; and (2) where a flash flood inundates a school site or cuts off evacuation access. In both these cases, advance planning should determine which areas of the facility are the best areas for temporarily sheltering people, and logistics for moving people quickly should be understood and exercised periodically.

F.5 Planning the Recovery

The first order of business after a flood is to determine if it is safe to reach the facility (e.g., are roads blocked or washed out, are power lines down, or are there gas line leaks?). Coordination with emergency management officials will be needed.

Once access is possible, the next task is to have an engineer or building professional assess the facility from the outside, generally characterize damage or dangerous conditions, and determine if the facility is safe to re-enter and reoccupy. ATC-45, *Field Manual: Safety Evaluation of Buildings after Windstorms and Floods* (ATC, 2004), or similar documents and protocols will be useful. School districts may wish to have engineers under contract in advance so safety evaluations can be done quickly. It is important to note that safety evaluations are not the same as a property damage evaluation.

Once it is safe to access the facility, there are two priorities: (1) documenting immediate post-flood facility condition and damage (photos, videos, and field notes should be made); and (2) cleaning and drying of the facility. Documentation will be needed for insurance and possible disaster assistance. Chapter 5 provides more detailed information on this topic.

Flood-damaged contents and nonstructural elements should be removed. Standing water should be pumped out, but caution should be taken if basements are flooded as rapidly pumping water out of basements with a high groundwater table could lead to basement wall or floor failure. A structural engineer should be consulted if there are any questions about pumping out basements. Commercial fans, dryers, and dehumidifiers should be brought in. Emergency generators may be required. Failure to properly and completely clean a flooded facility may lead to more problems later; most concerning would be mold contamination and possible health impacts to students and staff.

Once a facility is dried and cleaned, attention can shift to making repairs to the facility. Repairs could be minor or major, depending on the nature and extent of flood damage, and decisions will have to be made regarding whether to reoccupy the facility or to move classes and activities to other facilities.

Throughout the recovery period, documentation of activities and expenses should be compiled. Frequent coordination with insurers, lenders, and funding authorities may be required. Chapter 5 provides more detailed information on this topic.

Recovery plans should also consider impacts on school operations after a flood. For example, the following should be considered:

- How long might a school be out of service if it is flooded or inaccessible? Are there other locations where classes and educational activities could be held while the school is out of service?

- Does the school host civic or community activities, and what would the impact of school closure be on those?

- If a school is designated to be used as a temporary post-flood shelter, how will this impact school operations and recovery following an event?

F.6   Recommended Resources

Full citations of all references used to develop this supplement are listed in the References section in this *Guide*. The following is a list of recommended resources that might be useful for school leaders that are addressing school flood risk. In some cases, a document is both in the References section and listed here as a recommended resource.

ATC-45, *Field Manual: Safety Evaluation of Buildings After Wind Storms and Floods* (ATC, 2004). This document provides guidelines for evaluating whether it is safe to enter and reoccupy buildings damaged by flood or wind. It is intended to be used by designers and building professionals performing safety evaluations. This document does not provide guidance on determining damage and repair costs.

Children and Young People 'After the Rain has Gone' – Learning Lessons for Flood Recovery and Resilience (Walker et al., 2010). This report provides the findings from a participatory research project that was undertaken to identify key issues for children and young people related to flood events and the ensuing recovery process. This project—'Children, Flood and Urban Resilience: Understanding children and young people's experience and agency in the flood recovery process'—was conducted in the United Kingdom. http://eprints.lancs.ac.uk/49462/1/FINAL_REPORT.pdf

FEMA P-424, *Risk Management Series: Design Guide for Improving School Safety in Earthquakes, Floods, and High Winds* (FEMA, 2010a). This document is primarily intended for architects and engineers. It primarily pertains to new schools, but does provide guidance for existing schools. https://www.fema.gov/media-library/assets/documents/5264

FEMA P-758, *Substantial Improvement/Substantial Damage Desk Reference* (FEMA, 2010b). This document provides practical guidance and suggested procedures relevant to the National Flood Insurance Program (NFIP), and

more information on substantial improvement/substantial damage.
https://www.fema.gov/media-library/assets/documents/18562

Floodsite. A European Union scientific project on floods created a
comprehensive website for secondary school students and teachers to learn
about floods. The website includes a variety of lessons and activities that can
be used as-is, or as templates for new lessons and activities. www.floodsite
.net/juniorfloodsite/.

National Flood Insurance Program: Flood Hazard Mapping. This website
provides information on FEMA's flood hazard mapping program—Risk
Mapping, Assessment and Planning (Risk MAP). Through this program,
FEMA identifies flood hazards, assesses flood risks and partners with states
and communities to provide accurate flood hazard and risk data to guide
them to mitigation actions. https://www.fema.gov/national-flood-insurance
-program-flood-hazard-mapping# and http://www.floodsmart.gov.

Risk Mapping, Assessment and Planning (Risk MAP), FEMA. This page
discusses the Risk MAP program and what the program can mean to
communities. This page is intended for a variety of audiences, including
state and community officials; homeowners, renters and business owners;
real estate, lending, insurance professionals; engineers, surveyors and
architects. https://www.fema.gov/risk-mapping-assessment-and-planning
-risk-map

So, You Live Behind a Levee! (ASCE, 2009). This book was developed to
help provide more information about levees and their associated risk. In
particular, it provides actionable advice to better protect users against future
flood hazards. http://ascelibrary.org/doi/pdf/10.1061/9780784410837

FEMA P-1000   **F: Floods**   **F-17**

Supplement H

Hurricanes

This supplement is applicable to all schools located in hurricane hazard areas. For schools that are located in both a hurricane and tornado-prone region, the *Tornado Supplement* is also applicable. The storm surge portion of this supplement is applicable to a relatively narrow strip of land (usually tens of miles, but sometimes greater) inland of ocean, bay, and tidal river shorelines along the Atlantic and Gulf of Mexico coasts. If a school site is vulnerable to storm surge or flood, the *Floods Supplement* and the *Flood Appendix* are also applicable.

"Storm surge" is the abnormal rise in coastal water level due to high winds and low atmospheric pressure, over and above the predicted astronomical tide.

This supplement provides hurricane wind and storm surge guidance for existing school buildings and guidance for new schools that are in the planning stage. The guidance is based on field observation research conducted on a large number of schools (and other buildings) that were affected by hurricanes.

After reading this supplement, school administrators, school emergency managers, teachers, and other school leaders should be able to:

- Determine if their school is located in a hurricane hazard area;

- Recognize the importance of having hurricane wind and storm surge vulnerability assessments performed for existing schools;

- Recognize the importance of siting new schools outside areas vulnerable to storm surge or flood;

- Recognize the importance of incorporating special design enhancements in new facilities to achieve greater wind resilience, and where necessary, storm surge and flood resilience;

- Create or update a school disaster plan with specific considerations for hurricanes; and

- Identify aspects that should be considered to facilitate school recovery following a hurricane.

H.1 Overview of Hurricanes

A hurricane is a system of spiraling winds converging with increasing speed toward the storm's center (the eye of the hurricane). Hurricanes form over warm (tropical) ocean waters. Typical hurricanes are about 300 miles wide, although they can vary considerably. The largest one on record had a diameter of 1,350 miles and the smallest was 60 miles. Hurricanes can also cause high storm surge, usually limited to less than 100 miles of coastline (see Figure H-1), and high waves and erosion that extend along hundreds of miles of coastline.

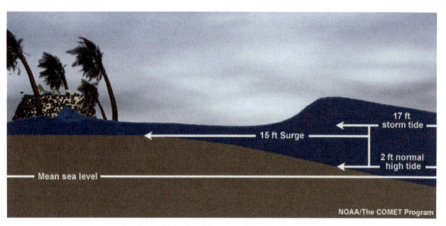

Figure H-1   Storm surge rises above the normal tide (National Hurricane Center, 2016).

> For more detailed information on the Saffir-Simpson Hurricane Wind Scale, visit www.nhc.noaa.gov/aboutsshws.php.

The Saffir-Simpson Hurricane Wind Scale categorizes the intensity of hurricanes based on wind speed, but does not include storm surge. The five-step scale ranges from Category 1 (the weakest) to Category 5 (the strongest). The terms "hurricane," "cyclone," and "typhoon" describe the same type of storm. The term used depends on the region of the world where the storm occurs.

In addition, being capable of delivering extremely strong winds for several hours and moderately strong winds for a day or more, many hurricanes also bring very heavy rainfall over areas extending from the coast to far inland, potentially causing severe flooding, and wind-driven rain damage to the interior of buildings that have been exposed due to damaging winds. Hurricanes also occasionally spawn tornadoes.

Hurricane season runs from June through November in the Central Pacific, Atlantic, Caribbean, and Gulf of Mexico regions. About six hurricanes can be expected during typical hurricane seasons and up to 15 hurricanes during more active years (NOAA, 2015). In the West Pacific region, hurricanes can occur in every season.

H.1.1 Hurricane Impacts on Schools

Of all the windstorm types, hurricanes have the greatest potential for devastating a large geographical area and, hence, affect the greatest number of schools during an event.

Because of the significant warning time for hurricanes, schools are typically not occupied during the event, unless they are used as a hurricane evacuation shelter. The focus of school building design for hurricanes is usually on minimizing building damage in order to facilitate reopening the school and to minimize repair cost. This is different from schools in tornado-prone regions where the school could be occupied during the tornado because of short warning. For those cases, the focus is on keeping the students safe during the tornado via the incorporation of tornado safe rooms/shelters.

> **The Children of Katrina**
>
> A seven-year study of students affected or displaced by 2005 Hurricane Katrina found that many New Orleans students suffered on personal, family, and social levels, and had declining educational trajectories (see Figure H-2). Many never recovered the former stability in their lives. Family disruption, loss of homes, displacement, and other factors all contributed to challenges of students concentrating in schools, higher anxiety levels, and more behavioral problems.
>
>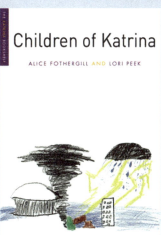
>
> Figure H-2   This book documented the many long-term, negative effects that Hurricane Katrina had on children (Fothergill and Peek, 2015).

Hurricane wind speeds are the greatest at the coast, but very high winds can occur inland for many miles before they decay significantly. The combination of strong and long-duration wind, wind-borne debris, and rain can severely damage schools, as shown in Figure H-3. Roof membrane blow-off or puncture by wind-borne debris typically results in water leakage, which can saturate fiberglass insulation and ceiling boards causing them to collapse, as shown in Figure H-4. Previous powerful hurricane winds have resulted in significant damage to a large number of schools in the impacted area.

Figure H-3   Large portions of the school roof coverings blew off (indicated by red arrows) during 2005 Hurricane Katrina (FEMA, 2010a).

Figure H-4   Damage to school ceiling in Mississippi during 2005 Hurricane Katrina (FEMA, 2006b).

Flooding from hurricanes can also cause severe damage to school facilities that are not elevated above the storm surge and any accompanying waves, and above flooding induced by heavy rains. Inundation by storm surge can cause damage to buildings and contents, as shown in Figures H-5 and H-6.

The effects of waves and high flow velocities close to an ocean shoreline can be more consequential. They usually cause more serious damage to the school facility (i.e., to the structural elements of the school). Strong waves can destroy many school buildings.

Figure H-5   Storm surge flooding in a high school in LaPlace, Louisiana during Hurricane Isaac in 2012. (Photo source: Rusty Costanza, *The Times-Picayune*)

Figure H-6   Storm surge flooding of a high school in Sabine Pass, Texas during Hurricane Ike in 2008. (Photo source: Jocelyn Augustino, FEMA)

Inland Rainfall from Hurricanes and Tropical Systems Can Cause Extensive Flooding

People know that hurricanes bring high winds and storm surge to the coast, but hurricanes and lesser tropical systems can lead to extreme flooding, even when high winds have abated. History is full of examples where tropical depressions, tropical storms, and remnants of hurricanes have flooded communities, including schools. The following is a selection of examples.

- The remnants of Hurricane Camille (1969, landfall in Mississippi) caused record flooding in central Virginia, where over 24 inches of rain fell in eight hours. Landslides, flash flooding, and river flooding followed the flooding (Romano, 2010).

- The remnants of Hurricane Agnes (1972, landfall in Florida) led to 7-15 inches of rain and widespread flooding across portions of Virginia, Maryland, and Pennsylvania (Weather Prediction Center, 2016a).

- Tropical Storm Allison (2001, landfall in Texas) led to widespread flooding in Houston and Harris County, which saw as much as 38 inches of rain in six days. Heavy rains continued across southeastern and mid-Atlantic states (Stewart, 2011).

- The remnants of Hurricane Frances (2004, landfall in Florida) led to over 20 inches of rain in western North Carolina, over 10 inches of rain in western Virginia, and 5-7 inches of rain in parts of Ohio, western Pennsylvania, and western New York (Weather Prediction Center, 2016b).

- An August 2016 stalled tropical depression dropped 12-30 inches of rain over the greater Baton Rouge area in three days (see Figure H-7). These record rains led to extensive flooding. Over 100,000 homes were flooded. School schedules in over 25 parishes were disrupted for one to four weeks, with 265,000 students affected. Flood damage affected homes of thousands of teachers and school staff. Hundreds of school buses were damaged by flooding. Ten schools were forced to temporarily relocate and school officials noted that some of these schools may never reopen. School damage repairs were still ongoing as of February 2017. Flood damages to schools are expected to total over $60 million (Lussier, 2017).

Figure H-7   Denham Springs, Louisiana, High School flooded in August 2016 (Photo source: Bill Feig, *The Advocate*).

Special attention to building siting, design, construction, and maintenance is needed to avoid prolonged disruption to school operations.

H.1.2 Improvements in School Construction for Hurricanes

Since the late 1980s, there has been significant improvement in understanding how to design and construct buildings to achieve good performance during hurricanes. Figure H-8 provides a synopsis of a selection of significant hurricane events and the improvement of building codes, test methods, and design guidelines. This has resulted in dramatically better performance of schools that were constructed after these improvements were implemented.

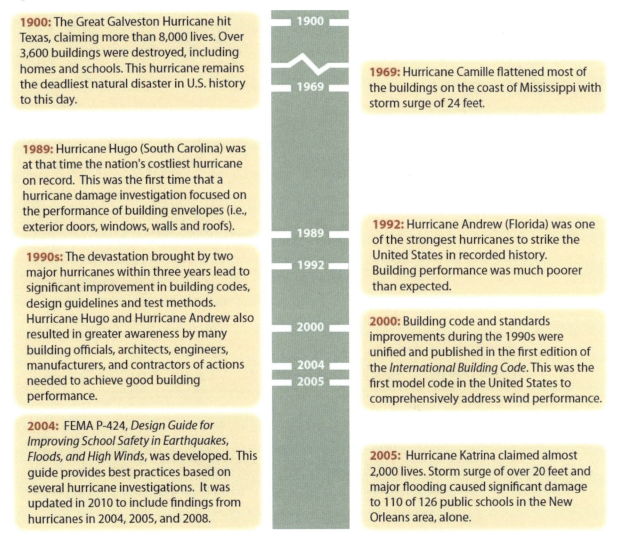

Figure H-8   Timeline indicating a selection of significant hurricane events and improvements to building codes and guidelines.

H.1.3 Important Terminology

Evacuation Shelter for Hurricanes. This refers to a building or portion thereof that is used during a hurricane to provide refuge for people who live in flood-prone areas or unusually weak housing. An evacuation shelter may or may not have been designed to be used as a refuge area from high winds. It is therefore recommended to not use a school as an evacuation shelter unless it was suitably designed and constructed for this purpose.

Hurricane Safe Room*. This refers to a building or portion thereof that has been designed for the purposes of providing protection from hurricanes (during the event) and complies with the hurricane provisions in FEMA P-361, *Safe Rooms for Tornadoes and Hurricanes: Guidance for Community and Residential Safe Rooms* (FEMA, 2015c). FEMA P-361 provides design and construction guidance for registered design professionals, as well as emergency management considerations.

> *All safe room criteria in FEMA P-361 are consistent with and, in some cases, are more conservative than the corresponding ICC 500 requirements. Safe rooms being constructed with FEMA grant funds must comply with ICC 500 and FEMA P-361.

Hurricane Shelter*. This refers to a building or portion thereof that has been designed for the purposes of providing protection from hurricanes (during the event) and complies with the hurricane provisions in ICC 500, *Standard for the Design and Construction of Storm Shelters* (ICC, 2014a).

Hurricane Recovery Shelter. This refers to a building or portion thereof that is used to provide emergency housing and other support services to the community in the days/weeks after a hurricane. This concept is referred to as an "emergency shelter" in the building codes, although it is not meant for buildings that would be used during an event. For buildings designated as emergency shelters per the building codes, the intent is to make them useable after an event and are therefore designed to higher loads than typical.

H.2 Is Your School in a Hurricane Hazard Area?

Figure H-9 shows the tracks of hurricanes that have struck the Atlantic and Gulf coastal areas and Hawaii. As shown in Figure H-10, hurricane hazard areas include Atlantic and Gulf coastal areas, Hawaii, and the United States territories in the Caribbean and South Pacific, including American Samoa, Guam, Northern Mariana Islands, Puerto Rico, and the U.S. Virgin Islands. Schools located in a hurricane hazard area should include hurricane risk in their school hazard safety strategies.

Figure H-10 shows two shaded areas—the red area is prone to both high winds and heavy rain caused by hurricanes and the orange area is mostly subject to hurricane-induced heavy rain although high winds can also occur

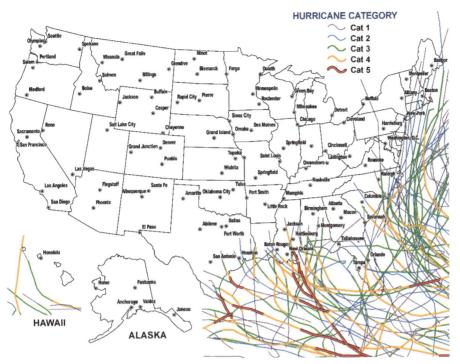

Figure H-9   Recorded Category 1–5 hurricanes striking the Atlantic and Gulf coastal areas and Hawaii from 1950 to 2014 (Source: Map data from NOAA, 2016a).

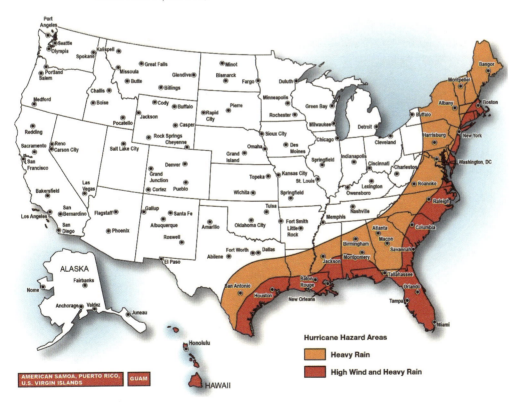

Figure H-10   Map indicating hurricane hazard areas (both orange and red zones) in the United States (high wind area adapted from ASCE, 2017b).

FEMA P-1000   H: Hurricanes   H-9

in those regions. The red zone is considered to have a significantly higher wind hazard and therefore, the building code specifies more stringent requirements for buildings, including schools, in this zone. To determine if a school is in the red shaded area of Figure H-10, enter the school's address at windspeed.atcouncil.org to obtain the design wind speed; the output shows design speeds for different risk categories. If the Risk Category II speed is greater than 115 miles per hour, the school is in the region that is prone to both high wind and heavy rain from hurricanes. Note that schools are predominately Risk Category III; however, the hurricane-prone region (red zone in Figure H-10) is defined by Risk Category II.

If a school is in a hurricane hazard area, it might also be affected by hurricane storm surge and other flooding. Several sources, noted below, should be referred to in order to determine if a school is in a storm surge area or flood hazard.

- Hurricane storm surge inundation areas have been mapped as part of hurricane evacuation studies for Atlantic and Gulf of Mexico states. Local or state emergency management agencies can usually provide this information. Figure H-11 provides an example of a hurricane storm surge inundation map.

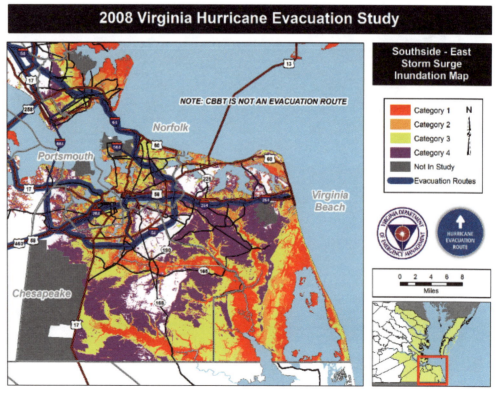

Figure H-11   Example hurricane storm surge inundation map in the Norfolk, Virginia area. Inundation areas are tied to Hurricane Categories (Virginia Department of Emergency Management, 2008).

- Hurricane storm surge depths above ground have been mapped by hurricane category for Atlantic and Gulf of Mexico states by the National Oceanic and Atmospheric Administration (NOAA) here: https://goo.gl/lpjFEQ.
- Flood Insurance Rate Maps (FIRMs) have been developed by FEMA for over 21,000 communities in the United States and its territories. Flood hazard maps, including FIRMs, are covered in detail in the *Floods Supplement*.

H.3 Making Buildings Safer

This section provides guidance on making existing and new school buildings safer from hurricane wind, storm surge, and rain hazards.

H.3.1 Existing School Buildings

Existing schools, particularly those constructed before 2000, are typically susceptible to significant wind damage as shown in Figure H-12, even during weak hurricanes. Figures H-13 and H-14 illustrate examples of other building vulnerabilities from hurricanes.

> FIRM Maps can be viewed and downloaded from FEMA's Map Service Center (MSC) https://msc.fema.gov/portal. Maps can be searched by street address or by state and community. Flood hazard information can also be viewed using the National Flood Hazard Layer, accessible via the MSC.

> FEMA P-424, *Risk Management Series: Design Guide for Improving School Safety in Earthquakes, Floods, and High Winds* (FEMA, 2010a), provides vulnerability assessment guidance to architects and engineers, and provides examples of retrofit measures that are often applicable.

Figure H-12 Damage in a school in Florida during Hurricane Andrew in 1992. Several windows in this school were broken by wind-borne debris, and a large portion of the roof membrane blew off. Resulting interior water damage precluded quick resumption of school. (Photo source: Thomas Smith)

Figure H-13   Collapsed unreinforced masonry classroom wall in the U.S. Virgin Islands during Hurricane Marilyn in 1995 (FEMA, 2010a).

Figure H-14   Rooftop mechanical equipment is often inadequately attached (even on new schools). Several exhaust fans blew off their curbs, allowing rain to directly enter the school during Hurricane Andrew, Florida in 1992. It is very inexpensive to anchor equipment and avoid costly damage (FEMA, 2010a).

To address these vulnerabilities, the following are recommended:

- **Wind Vulnerability Assessment.** It is recommended to have a wind vulnerability assessment performed by a qualified architectural and engineering team. The purpose is to identify and prioritize items that need to be strengthened or replaced. A proactive approach in mitigating weaknesses can avoid or decrease occupancy interruption and significantly reduce repair costs after a future hurricane. FEMA P-424, *Risk Management Series: Design Guide for Improving School Safety in Earthquakes, Floods, and High Winds* (FEMA, 2010a) provides vulnerability assessment for architects and engineers, as well as examples of mitigation (retrofit) measures that are often applicable.

- **Storm Surge Considerations.** If a school is located in an area that could be affected by hurricane storm surge, Section F.3.1 in the *Floods Supplement* applies. That section provides information regarding vulnerability assessments and making existing school buildings safer from storm surge.

- **Prioritizing Remedial Work.** Before beginning remedial work, it is necessary to understand all significant aspects of a school's vulnerability with respect to wind, storm surge, and wind-driven rain. If funds are not available to correct all identified deficiencies, the work should be systematically prioritized so that the items of greatest need are corrected first. Mitigation efforts can be very ineffective if they do not address all significant items that are likely to fail. For example, if a school has windows that are vulnerable to breakage from wind-borne debris and also has a roof system that is vulnerable to wind blow-off, both should be mitigated. If installation of window shutters is the only mitigation effort taken, blow-off of the roof would likely still result in significant interior damage due to water leakage.

- **Planned Repairs.** It is recommended that mitigation work be incorporated when performing planned repairs or replacements. For example, when replacing a roof covering that has reached the end of its service life or been damaged by hail, the roofing recommendations provided in FEMA P-424 can be incorporated. However, if a roof system is extremely susceptible to wind damage, it should be mitigated as soon as budget permits. Repairing hurricane damage also presents an opportunity to incorporate FEMA P-424 recommendations.

- **Window Protection.** The 2000 and subsequent editions of the *International Building Code* require that exterior windows be protected from wind-borne debris in portions of hurricane-prone regions (i.e., the "wind-borne debris region"). Most schools built prior to 2000 have windows that can easily be broken by wind-borne debris. FEMA P-424 recommends window protection in a larger geographical area than required by the code. For schools located in the area prescribed in FEMA P-424, window mitigation can be economically implemented by installing shutters, as shown in Figure H-15.

- **Maintenance.** The exterior of the building should be inspected yearly by maintenance personnel. Items that have deteriorated (such as leaky roofs) or become loose (such as rooftop equipment) should be repaired or replaced before they are damaged by wind.

- **Removal of Nearby Potential Hazards.** Per FEMA P-424 recommendations, trees with trunks larger than six inches in diameter should be moved away from the school so that they do not cause damage if they topple. Figure H-16 shows why this is important.

Figure H-15 This school was retrofitted with accordion shutters prior to a hurricane. The shutters successfully protected the windows from wind-borne debris during 2004 Hurricane Ivan (FEMA, 2010a).

Figure H-16 Had this tree fallen in the opposite direction during 2004 Hurricane Ivan in Florida, it would have damaged the school (FEMA, 2007).

> **Making Existing School Buildings Safer: Summary of Key Points**
>
> - Determine if your school is in a hurricane hazard area.
>
> - Have a vulnerability assessment performed by a qualified architectural and engineering team. If your school is also in a storm surge or rainfall flood hazard area, make sure that the vulnerability assessment also includes flood vulnerabilities.
>
> - Mitigate significant wind vulnerabilities.
>
> - Mitigate storm surge and flood vulnerabilities where feasible; consider building elevation or relocation, or decommissioning and replacement with a new school facility outside the storm surge-prone region.

H.3.2 New School Buildings

Building codes, standards, and design guidelines have greatly improved since devastating hurricanes in 1989 and 1992, as shown Figure H-8. However, complying with the *2015 International Building Code* (ICC, 2014b) does not ensure that a school will be operational after an event. Portions of the code are not conservative, such as the extent of the wind-borne debris region. Due to documentation of broken windows outside of the wind-borne debris region prescribed by the code, FEMA P-424 recommends window protection further inland than required by the *2015 International Building Code*. Additionally, the code does not address some important wind performance issues, such as wind-resistance of roof gutters and water leakage caused by wind-borne debris puncture of roof systems. To avoid prolonged occupancy interruption and to minimize the cost of building damage, the special design enhancements (best practices) in FEMA P-424 are recommended. The increased cost to implement the best practices is estimated to be less than 3% for new construction.

For schools located in hurricane storm surge areas, considerations outlined in Sections F.3.2 in the *Floods Supplement* should be incorporated. Those sections provide information regarding important siting and design considerations for new school buildings potentially subject to flooding.

H.3.2.1 Shelters During the Event

Hurricane evacuation shelters provide refuge for people who live in flood-prone areas or unusually weak housing. Local, county, and school emergency managers should be consulted during the planning stage for a new school to determine if it will be used as an evacuation shelter. If it will be used as a shelter during a hurricane, it should be designed in accordance

---

FEMA P-1000   **H: Hurricanes**   **H-15**

with the hurricane shelter criteria in ICC 500 and FEMA P-361. These criteria are more stringent than for normal school buildings. For example, shelters should be designed for higher wind speeds and should not be located in flood-prone areas. FEMA P-361 indicates that the increased cost to comply with the hurricane safe room criteria is estimated to be less than 5% for new construction.

FEMA P-361 also has criteria for tornado safe rooms/shelters. The hurricane safe room/shelter criteria are different from that of tornadoes.

H.3.2.2 Shelters After the Event

Hurricane recovery shelters provide emergency housing and other support services to the community in the days/weeks after an event. School, local, and county emergency managers should be consulted to determine if a school is intended to be used or designated as a recovery shelter. If so, it should be classified as a Risk Category IV building per the building codes, requiring stronger design loads. However, shelter design goes far beyond a stronger building and should also accommodate prolonged interruption of electrical power, sewer, and water service, and other features desired by the local emergency services agency. Such provisions in schools will also significantly improve performance and minimize school closures due to damage (SEFT Consulting Group, 2015).

In many cases, schools are designated to be emergency shelters without first making sure that they are designed to be functional following a hurricane. Figure H-17 shows an example of school buildings that were too damaged to be used as recovery shelters following a hurricane. FEMA P-424 and FEMA P-1019, *Emergency Power Systems for Critical Facilities: A Best Practices Approach to Improving Reliability* (FEMA, 2014), provide relevant design guidance. Alternatively, advanced arrangements could be made to provide delivery of an emergency generator and water after the event, prior to opening of the center.

Making New School Buildings Safer: Summary of Key Points

- Determine if your school is in a hurricane hazard area.

- Determine if the school will be used as a shelter during the event.

- Determine if the school will be used as a shelter after the event.

- Have the architectural and engineering team implement the guidance, as relevant, in Section H.3.2.

Figure H-17   View of aid tents between a damaged elementary school (red arrow) and high school (foreground) in Florida after Hurricane Andrew. (Photo source: Thomas Smith)

H.4   Planning the Response

Hurricane hazard areas require careful emergency planning given that hurricanes can require massive evacuations over a short period of time, they can cause widespread devastation, and they can affect many schools in one event. These should be major considerations in school emergency operations plans.

School buildings are often designated as hurricane evacuation shelters even though they were not designed and constructed for this purpose. In many instances, designated buildings did not perform well during hurricanes, and have sometimes been evacuated during the event. In addition to shelter occupants being at risk of injury, they may have a very negative experience, such as getting wet from roof leakage or back-up of toilets. If they return to a home that experienced little damage, they may be reluctant to evacuate in a future event that could destroy their home. It is therefore recommended to not use a school as an evacuation shelter unless it was suitably designed and constructed for this purpose.

Emergency operations plans should indicate how to prepare the building if a hurricane is forecast. For example, shutters should be closed, the roof and grounds should be checked, and items that could become wind-borne debris should removed or secured. School management should confirm with the pre-contracted architects, engineers, and contractors that they will immediately assess the school and perform emergency repairs.

> In other types of windstorms, students might not be able to evacuate in advance as with hurricanes. For guidance on what to do in those situations, see the *High Winds Supplement*.

For schools that might also be affected by hurricane storm surge or flooding, the National Hurricane Center may issue storm surge inundation/depth maps to show anticipated storm surge flooding for a hurricane within 48 hours of expected landfall (see http://www.nhc.noaa.gov/surge/inundation/). Section F.4 in the *Floods Supplement* provides additional information and guidance.

Flood-Resistant Design Passes a Test in Port Bolivar, Texas

Figure H-18 shows an elementary and middle school building constructed in 2005 on the Bolivar Peninsula, just east of Galveston, Texas. Hurricane Ike made landfall nearby in September 2008, and design flood conditions affected the site. The building sustained some minor wind damage, but no flood damage. The building was elevated on concrete columns approximately one story higher than required by the flood map and building code. Flooding during Hurricane Ike reached approximately 5.5 ft above the ground—approximately design flood conditions—but did not enter the building. The building was one of the few buildings standing after Hurricane Ike and served many purposes after the storm. It was used to house emergency responders and to host many community meetings during the recovery and reconstruction period. Students returned in February 2009. FEMA (2009a) and Powitzky (2009) provide more information about the school and its operations after Hurricane Ike.

Figure H-18   Crenshaw Elementary and Middle School, located in Port Bolivar, Texas, which was successfully designed to be a hurricane shelter in 2005 and was one of the few buildings standing after 2008 Hurricane Ike. (Photo source: Laurie Johnson)

H.5 Planning the Recovery

As mentioned in Section H.4, hurricanes can require massive evacuations over a short period of time. Recovery plans should consider that hurricanes can devastate a large geographical area and can impact many schools during a single event. Recovery plans should also consider the impact of local evacuation routes on school closures and the availability of school staff.

Schools affected by a hurricane should be evaluated by an architectural and engineering team to determine whether or not they are safe to reoccupy.

ATC-45, *Field Manual: Safety Evaluation of Buildings After Wind Storms and Floods* (ATC, 2004), provides guidance on rating the safety significance of damage. The architectural and engineering team should also determine whether there was damage that is not directly related to life safety. Having pre-event contracts with architectural and engineering teams, and with contractors (including roofing contractors) for immediate post-event damage assessment and emergency repairs is highly recommended.

If the school will be used as a shelter either during or after the event (see Sections H.3.2.1 and H.3.2.2, respectively), additional aspects should be considered when planning for recovery. These include considering the time and expenses required to clean the site prior to resumption of school, and contingency plans to use an alternate building in case damage precludes it from being used as such.

Even if the school is expected to have only minor or no damage, school might not be able to resume for a week or two. Time may be needed for clearing of roads, confirmation that the school is safe to reoccupy, and restoration of basic services, such as electrical power, sewer, and water.

If there is significant damage, an evaluation should be performed to determine if it is cost-effective to repair the damage versus demolish the building and build a new one. If the damage is severe enough, repairs will need to include various code-required upgrades (such as fire alarm systems), which can dramatically increase cost and may economically necessitate replacement with a new building, such as in the example shown in Figure H-19.

Figure H-19   Essentially the entire roof covering on this older school blew off (Typhoon Paka, Guam, 1997). It is often more economical to demolish an old school such as this and build a new school, rather than perform repairs and required upgrades.

If new construction is necessary, school leaders should consider implementing the guidance provided in Section H.3.2. If the damage is repairable, in addition to performing repairs, it is recommended that a vulnerability assessment be performed as described in Section H.3.1.

For schools that might also be affected by hurricane storm surge or flooding, Section F.5 in the *Floods Supplement* provides recovery considerations and guidance specific to flooding and storm surge.

H.6   Recommended Resources

Full citations of all references used to develop this supplement are listed in the References section in this *Guide*. The following is a list of recommended resources that might be useful for school leaders that are addressing school hurricane risk. In some cases, a document is both in the References section and listed here as a recommended resource.

ASCE/SEI 24, *Flood Resistant Design and Construction* (ASCE, 2014). This document sets forth the minimum design and construction requirements for buildings in flood hazard areas.

ATC-45, *Field Manual: Safety Evaluation of Buildings After Wind Storms and Floods* (ATC, 2004). This document provides guidelines for evaluating whether it is safe to enter and reoccupy buildings damaged by wind or flood. It is intended to be used by designers and building professionals performing safety evaluations. This document does not provide guidance on determining damage and repair costs.

FEMA P-361, *Safe Rooms for Tornadoes and Hurricanes: Guidance for Community and Residential Safe Rooms* (FEMA, 2015c). This document is primarily intended for architects and engineers; however, it provides emergency management considerations for the operations and maintenance of tornado and hurricane safe rooms. https://www.fema.gov/media-library/assets/documents/3140

FEMA P-424, *Risk Management Series: Design Guide for Improving School Safety in Earthquakes, Floods, and High Winds* (FEMA, 2010a). This document is primarily intended for architects and engineers. It primarily pertains to new schools, but does provide guidance for existing schools. https://www.fema.gov/media-library-data/20130726-1531-20490-0438/fema424_web.pdf

FEMA 488, *Mitigation Assessment Team Report: Hurricane Charley in Florida: Observations, Recommendations, and Technical Guidance* (FEMA, 2005a). This document provides information about Hurricane

Charlie effects on Florida buildings, including school buildings. FEMA has produced similar documents after other major hurricanes. http://www.fema .gov/media-library-data/20130726-1445-20490-6387/fema488.pdf

FEMA 549, *Hurricane Katrina in the Gulf Coast: Building Performance Observations, Recommendations, and Technical Guidance* (FEMA, 2006b). This document provides information about Hurricane Katrina effects on the Gulf Coast, including school buildings. FEMA has produced similar documents after other major hurricanes. http://www.fema.gov/media -library/assets/documents/4069

FEMA P-757, *Mitigation Assessment Team Report, Hurricane Ike in Texas and Louisiana: Building Performance Observations, Recommendations, and Technical Guidance* (FEMA, 2009a). http://www.fema.gov/media-library /assets/documents/15498. This document provides information about Hurricane Ike effects on Texas and Louisiana buildings, including school buildings. FEMA has produced similar documents after other major hurricanes. See http://www.fema.gov/fema-mitigation-assessment-team -mat-reports.

FEMA P-1019, *Emergency Power Systems for Critical Facilities: A Best Practices Approach to Improving Reliability* (FEMA, 2014). This document is primarily intended for architects and engineers, https://www .fema.gov/media-library/assets/documents/101996

Hurricane Ready.gov website. This webpage provides guidance on what actions to take during a hurricane watch or warning alert from the National Weather Service. It also provides tips on what to do before, during, and after a hurricane. https://www.ready.gov/hurricanes

ICC 500, *Standard for the Design and Construction of Storm Shelters* (ICC, 2014a). This document sets forth requirements for design and construction of buildings or portions thereof to be used as storm shelters.

Introduction to Storm Surge (NOAA). This document provides basic information on storm surge, how it is measured, and its effects. http://www .nhc.noaa.gov/surge/surge_intro.pdf

JetStream – An Online School for Weather. This website was developed to help educators, emergency managers, or any other interested party in learning about weather and weather safety. http://www.srh.noaa.gov /jetstream/index.html

National Hurricane Center's Hurricane Preparedness webpage. This webpage provides guidance on hurricane preparedness, as well as information on Hurricane Preparedness Week. http://www.nhc.noaa.gov /prepare/ready.php

The Storm as a Teacher: Lessons in Preparedness from Hurricanes Ike and Rita (Powitzky, 2009). This document describes how a new school building was designed to resist hurricane wind and surge affects, and served as a recovery and community center after Hurricane Ike.

The StormReady® Program, hosted by the National Weather Service, offers a community recognition program for communities and institutions that implement storm ready activities, including activities specific to preparing for hurricanes. To learn how to become StormReady®, visit http://www .weather.gov/stormready/

Storm Surge Fast Draw. This short video provides an explanation of storm surge and its effects. https://www.youtube.com/watch?v=bBa9bVYKLP0

Supplement TO

Tornadoes

Winds with sufficient speed to cause building damage can occur anywhere in the United States and its territories. However, tornadoes and hurricanes present the greatest hazard, and tornadoes are the deadliest type of windstorm. This supplement is applicable to all schools located in tornado-prone regions. Additionally, the *High Winds Supplement* provides guidance for other types of high winds. For schools that are located in both a tornado- and hurricane-prone region, the *Hurricanes Supplement* is also applicable.

This supplement provides occupant protection guidance for existing buildings and for new schools that are in the planning stage. The guidance is based on field observation research conducted on a large number of schools (and other buildings) that were affected by tornadoes.

After reading this supplement, school administrators, school emergency managers, teachers, and other school leaders should be able to:

- Determine if a school is in a tornado-prone region;
- Determine geographical areas where tornado safe rooms/shelters are required, and where they are recommended;
- Determine if an architect or engineer should be retained to identify best available tornado refuge areas;
- Create or update a school disaster plan with specific considerations for tornadoes; and
- Identify aspects that should be considered to facilitate school recovery following a tornado.

TO.1 Overview of Tornadoes

A tornado is a violently rotating column of air extending from the base of a thunderstorm to or near the ground. Tornado path widths are typically less than 1,000 feet, however, widths of approximately 2.5 miles have been reported. The National Weather Service (NWS) rates tornado severity according to the six levels of observed damage in the Enhanced Fujita Scale (EF Scale). The scale ranges from EF0 to EF5, as shown in Figure TO-1.

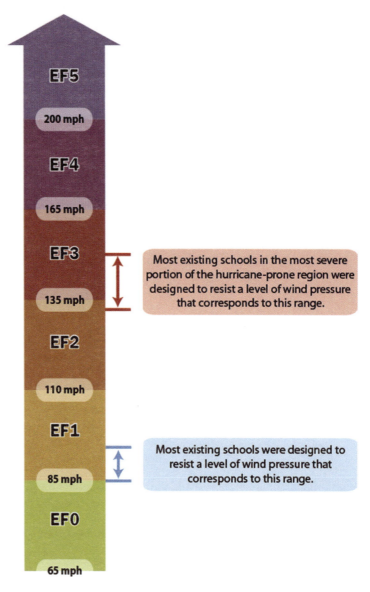

Figure TO-1   Enhanced Fujita scale (National Oceanic and Atmospheric Administration).

TO.1.1 Tornado Impacts on Schools

Figure TO-2 illustrates the type and increasing severity of school damage as a function of the EF rating. Most tornadoes are EF0 or EF1. These weak tornadoes can damage most schools, but they present a low risk to life safety. Approximately 5% of the tornadoes are EF3 or stronger. As shown in Figure TO-3, these tornadoes can cause significant damage to most schools, even if they are well designed, constructed, and properly maintained.

Tornadoes generate a large quantity of wind-borne debris, as illustrated in Figures TO-3 and TO-4. As the tornado intensity increases, the quantity of debris and risk of injury and death increase.

> Tornadoes challenge affected communities in many ways, but the most devastating effect is measured in deaths and injuries. The National Weather Service started keeping organized records of tornadoes in the United States in 1950. Since then, the deadliest year for tornadoes was 2011, which claimed 553 lives (FEMA, 2015c).

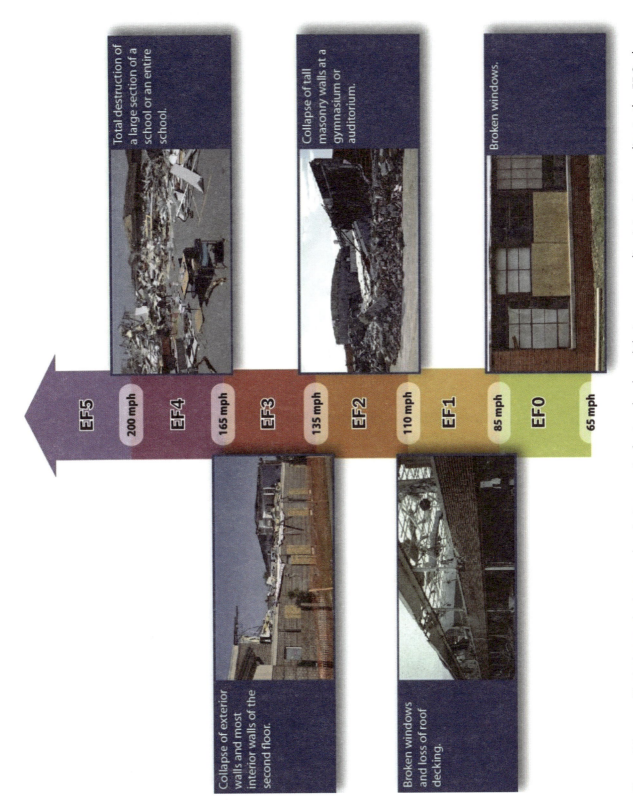

Figure TO-2　Typical tornado damage descriptions particular to schools and their corresponding intensity according to the EF Scale.

Figure TO-3   Aerial view of a school devastated by a strong tornado in Moore, Oklahoma in 2013 (Photo source: AP Images). Yellow circle indicates where a wind-blown car impacted an exterior wall (yellow line) and pushed it into a classroom. Red circle marks an upside down car in the hallway, and the red arrow indicates collapsed interior masonry walls.

Figure TO-4   Wind-borne debris in a classroom in Greensburg, Kansas in 2007. Note that all of the exterior windows are broken and the roof deck was blown off. (Photo source: Thomas Smith, FEMA)

Because of the extremely high wind pressures, the dangers of wind-borne debris, and low probability of occurrence at a specific location, tornado wind design philosophy for schools focuses on life safety, rather than on minimizing building damage.

TO.1.2 Important Terminology

Best Available Refuge Areas. This refers to areas in an existing building that have been deemed by a qualified architect or engineer to likely offer the greatest safety for building occupants during a tornado, as defined by FEMA P-431, *Tornado Protection: Selecting Refuge Areas in Buildings* (FEMA, 2009b). It is important to note that, because these areas were not specifically designed as a tornado safe room/shelter, their occupants may be injured or killed during a tornado. However, occupants in a best available refuge area are less likely to be injured or killed than people in other areas of a building. Best available refuge areas should be regarded as an interim measure only until a safe room/shelter is made available to the building occupants.

Tornado Safe Room. This refers to a building or portion thereof that complies with the tornado provisions in FEMA P-361, *Safe Rooms for Tornadoes and Hurricanes: Guidance for Community and Residential Safe Rooms* (FEMA, 2015c). FEMA P-361 provides design and construction guidance for registered design professionals, as well as emergency management considerations for the operations and maintenance of tornado safe rooms.

Tornado Shelter. This refers to a building or portion thereof that complies with the tornado provisions in ICC 500, *Standard for the Design and Construction of Storm Shelters* (ICC, 2014a).

Tornado Watch. The National Weather Service issues a Tornado Watch when conditions are favorable for tornado development.

Tornado Warning. The National Weather Service issues a Tornado Warning when a tornado has been sighted or indicated by weather radar.

TO.2 Is Your School in a Tornado-Prone Region?

Tornadoes can occur throughout the United States, as shown in Figure TO-5. However, the more destructive and deadly strong and violent tornadoes are rare in the west and northeast, as shown in Figure TO-6.

Figure TO-7 shows the design wind speeds that are used for tornado safe rooms/shelters. This map should be used to determine if a school is located in a tornado-prone region. Schools that are located in the 160, 200, and 250 miles per hour wind speed zones are considered to be in a tornado-prone

> Life Safety is an engineering term used to describe a level of design. The main goal behind life safety is to prevent fatalities and serious injuries in a building due to failure or collapse of structural elements, such as columns and beams.

> Safe rooms being constructed with FEMA grant funds must comply with ICC 500 and FEMA P-361. All safe room criteria in FEMA P-361 are consistent with and, in some cases, are more conservative than the corresponding ICC 500 requirements.

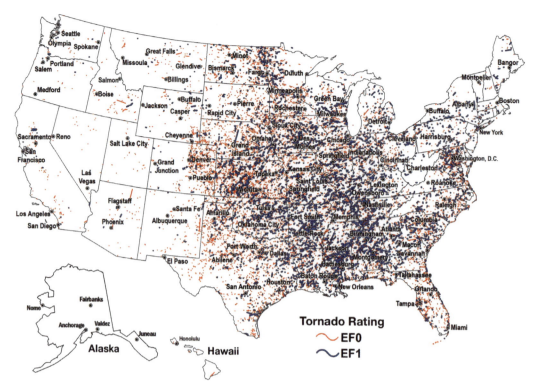

Figure TO-5   Recorded EF0 and EF1 tornadoes from 1950 to 2014. (Source: Map data from NOAA, 2016b)

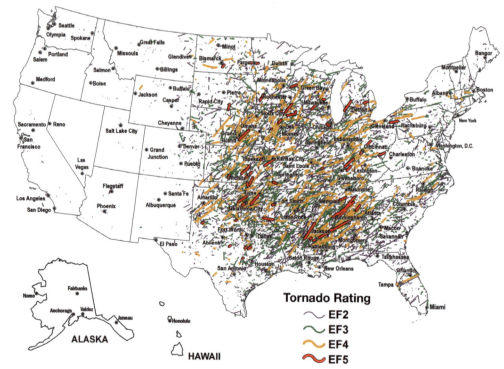

Figure TO-6   Recorded EF2, EF3, EF4, and EF5 tornadoes from 1950 to 2014. (Source: Map data from NOAA, 2016b)

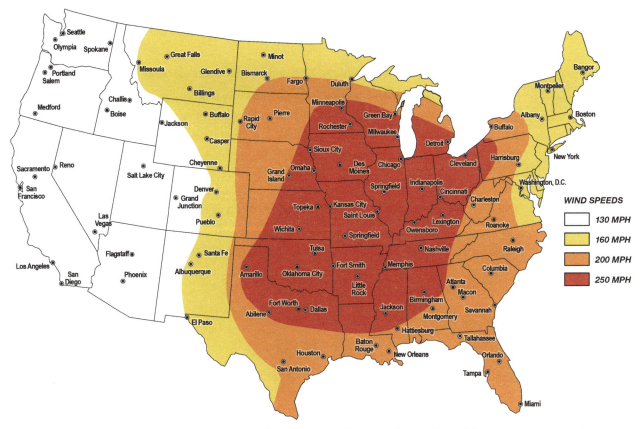

Figure TO-7　Safe room/shelter design wind speed zones for tornadoes (adapted from ICC, 2014a). The 2015 edition of the *International Building Code* requires new schools in the red zone to have a tornado shelter. FEMA recommends that schools in the yellow and orange zones also have a tornado safe room/shelter.

region. To determine if a school is in one of these three zones, see Figure TO-7. If a school is located along or near a boundary of different wind speeds on the Figure TO-7 map, the higher wind speed should be assumed.

TO.3　Protecting Occupants within School Buildings

This section provides guidance pertaining to occupant protection within school buildings.

TO.3.1　Existing School Buildings

For existing schools located in the 160, 200, or 250 mph wind zones, it is recommended that a safe room/shelter be constructed. Tornado safe rooms/shelters are specifically designed and constructed to provide occupant protection from high winds and wind-borne debris. The 2018 edition of the *International Existing Building Code* will, in certain circumstances, require ICC 500 compliant tornado shelters in K-12 school buildings with 50 or more occupants located in the 250 mph tornado wind speed zone shown in

> Tornado safe rooms/shelters are intended to provide occupant protection, and in addition, they can also reduce anxiety each time a tornado watch is issued. Students, faculty and staff in a school with a safe room/shelter should be comforted by knowing that there is an area in the school that will keep them safe. Peace of mind should facilitate learning while students are in class during a tornado watch.

Figure TO-7. A similar requirement is already in place for new school buildings (see Section TO.3.2).

A safe room/shelter can be sized to only accommodate students, faculty, staff, and visitors. It can also be sized to also accommodate nearby community members. When sized to also serve a community, operational procedures should be established to open the safe room/shelter during non-school hours. Section 3.2.1 and Section 3.2.2 provide more information on tornado safe room/shelter costs and timeline of development.

When a safe room/shelter is added to an existing campus, it is normally done as part of an addition, such as a classroom wing or multipurpose room as shown in Figure TO-8. In some instances, a portion of an existing school may be retrofitted to meet safe room/shelter criteria, however, this option is rarely cost effective.

Figure TO-8   This multipurpose room addition was designed as a safe room in Wichita, Kansas (FEMA, 2012d).

> Because best available refuge areas were not specifically designed to serve as safe rooms or shelters, their occupants may be injured or killed during a tornado. However, occupants in a best available refuge area are less likely to be injured or killed than people in other areas of a school.

As an interim until a safe room/shelter is available, it is recommended that best available refuge areas be identified by a qualified architect or engineer, using the guidance in FEMA P-431.

Many schools in the tornado-prone region have identified refuge areas. Often however, the areas were not selected by a qualified architect or engineer. When areas are selected by people with inadequate expertise, it may result in students, faculty, and visitors taking refuge in areas that offer less protection. The importance of having a qualified architect or engineer select the best available refuge areas is illustrated by the example from the Oklahoma training program, shown in Figure TO-9.

Determining Best Available Refuge Areas: The Value of Having a Trained Team

After deadly tornadoes struck Moore, Oklahoma in 2013, a training and school assessment program was conducted by FEMA and the State of Oklahoma. This included a two-day training program for architects, engineers, and building professionals on the design of tornado safe rooms and the selection of best available refuge areas. Teams were assigned an existing school and tasked with selecting best available refuge areas. The teams reviewed drawings of the existing buildings, conducted a site visit, analyzed findings, and recommended where to take refuge. This effort took less than eight hours with some additional time for report writing. Thus, the cost to have a qualified architect or engineer select refuge areas is minimal— around eight to sixteen hours for a small school and up to 24 hours for a large school.

Figure TO-9 shows the floor plan of a school indicating the previously identified refuge areas, as well as the best available refuge areas selected by the trained assessment team. Some unique construction characteristics of the building presented life safety hazards that were not recognized by the entity that previously selected the refuge areas. A comparison of previously identified refuge areas versus the best available refuge areas show a significant difference in the areas that were selected for refuge. While the space afforded by designated best available refuge areas is noticeably less than the space of the previously identified refuge areas, they were determined to be of sufficient size for the school's projected population needs.

Figure TO-9   Identified refuge areas, both previous and updated, marked on school floor plan. "Updated" refers to the best available refuge areas that were identified by the trained assessment team.

> *Avoid a false sense of protection. If a school is struck by a tornado and the refuge areas sustain little or no damage, it is important to avoid the conclusion that the refuge areas are safe. A future tornado that is stronger or on a slightly different track could destroy the refuge areas. Safe protection is only provided by a tornado safe room/shelter.*

> Building codes are adopted and can be modified at the local or state level. For example, as of September 2016, when the State of Oklahoma Uniform Building Code Commission adopted the 2015 edition of IBC, it removed the requirement for tornado shelters in new schools.
>
> In contrast, Alabama and Illinois both passed legislation in 2010 and 2015, respectively, requiring tornado shelters in all new schools. These statutes preclude local jurisdictions from deleting the tornado shelter requirement from the building code.

Signage Terminology. Refuge areas that do not comply with safe room/shelter criteria should not be posted with signage that would make occupants think they are in a safe room/shelter. It is recommended that best available refuge areas be posted with signs that say "Best Available Tornado Refuge Area."

TO.3.2 New School Buildings

Beginning with the 2015 edition of the *International Building Code*, ICC 500 compliant tornado shelters are required in new K-12 school buildings with 50 or more occupants located in the 250 mph tornado wind speed zone shown in Figure TO-7. New school buildings must abide by building code requirements that are adopted by local or State jurisdictions. The adopting entity can delete or modify the code provisions or add to them, unless precluded from doing so by the state statute. FEMA recommends that disaster-resistant code provisions not be weakened or omitted.

Whether or not required by the adopted building code in a particular area, it is recommended that a tornado safe room/shelter be incorporated in schools located in the 160, 200, and 250 mph wind speed zones.

TO.3.2.1 Tornado Safe Room/Shelter Cost

The cost to design and construct a new safe room versus normal construction within the 250 mph wind zone is estimated to be 20% to 32% greater. This increase only applies to the new safe room construction and not to the entire building. For example, if a new school's multipurpose room will also serve as a safe room, only the cost of the multipurpose room is estimated to increase by 20% to 32%. The cost of the remainder of the school is unchanged. The percent increase in cost for safe rooms in the 200 and 160 mph wind zones are lower. FEMA P-361 provides more detailed information on safe room costs and considerations.

TO.3.2.2 Tornado Safe Room/Shelter Timeline

Figure TO-10 provides a synopsis of major events related to development of safe room/shelter design criteria and building code adoption. Criteria for design and construction of tornado safe rooms/shelters were published in 2000, and new schools were first required by a building code to have a safe room/shelter in 2015.

1970s – 1990s: Texas Tech University's post-tornado damage observations led to the development of strategies for designing and testing of rooms within buildings that would protect occupants.

1970: Lubbock, Texas was struck by a violent tornado that caused 26 fatalities, over 1,500 injuries and extensive property damage. Faculty and students in Texas Tech University's civil engineering department investigated the event and pioneered field research investigations of subsequent tornadoes and hurricanes.

1999: A tornado outbreak in Kansas and Oklahoma caused 49 fatalities, approximately 800 injuries and extensive property damage. FEMA deployed a team to assess the damage. That investigation led to the development of FEMA 361, *Design and Construction Guidance for Community Shelters* (FEMA, 2000a). This publication provided architects and engineers reliable criteria for designing tornado safe rooms within buildings.

2007: Enterprise High School in Alabama was hit by an EF4 tornado, causing the deaths of 8 students. This tragedy led Alabama state legislators to enact a law that required all new public schools to incorporate tornado shelters.

2013: An Oklahoma tornado claimed the lives of 7 schoolchildren at Plaza Towers Elementary School. Several more students and teachers were injured in this and other buildings on the same campus. This same tornado also destroyed another elementary school, injuring several people, and collapsed the gymnasium at a high school.

2018: The *International Existing Building Code* will require additions to existing schools in the 250 mile per hour tornado wind region to have a tornado shelter, dependent on the size of the addition.

2011: A tornado outbreak in the Southeast caused 361 fatalities in April, and a tornado in Joplin, Missouri caused 161 fatalities in May. FEMA teams that investigated these events recommended that a building code change proposal be submitted to require new schools located in the 250 mile per hour tornado wind region to have a tornado safe room/shelter.

2015: The *International Building Code* started requiring new schools in the 250 mile per hour tornado wind region to have a tornado shelter.

Timeline markers: 1970, 1990, 1999, 2007, 2011, 2013, 2015, 2018

Figure TO-10   Timeline of development and requirements of tornado safe rooms/shelters.

Schools and the Joplin Tornadoes of May 2011

An EF5 tornado struck Joplin, Missouri in May 2011, causing 161 fatalities. It severely damaged five public schools and one private school. Fortunately, it struck on Sunday and the schools were not occupied. Five schools were replaced with new buildings and one school required major renovation work. The renovated school opened in fall of 2013, four of the new schools opened in 2014, and the private school opened in 2015. All of the schools incorporated a tornado safe room in the reconstruction.

The rebuilding of the public schools was made possible by passage of a bond measure, along with funds from building insurance and a FEMA grant to cover a significant portion of the safe room costs. The school district would not have been able to afford rebuilding one of the schools, or any of the safe rooms if the bond had not passed. The bond measure passed by a margin of less than 1%.

Joplin East Middle School was one of the damaged schools. At the gymnasium, two roof trusses and an exterior wall collapsed, as shown in Figure TO-11. At the auditorium, two exterior walls collapsed. The poor performance was not due to building age (it was constructed in 2009). Rather, the damage was caused by wind speeds that substantially exceeded those that the school was designed to withstand.

This school had six interior rooms designated as areas of "Tornado Safe Shelter;" however, they did not possess the wind pressure and wind-borne debris resistance required for adequate tornado safe rooms/shelters.

An industrial park warehouse was converted into a temporary school for the start of the 2011-2012 school year, and was used until the new school opened in January 2014.

Figure TO-11   View of the gymnasium. A brick veneer/precast concrete wall and two roof trusses collapsed onto the floor (FEMA, 2012d).

Protecting Occupants: Summary of Key Points

- Determine if your school is in a tornado-prone region.

- A tornado shelter is required for new schools constructed in the 250 mph wind zone.

- A tornado safe room/shelter is recommended for new schools and school additions in the 160 mph and greater wind zones.

- For schools located in the 160, 200, and 250 mph wind zones that do not have access to safe rooms/shelters:

  o Plan for the construction of a safe room/shelter.

  o In the interim, have a qualified architect or engineer select best available refuge areas.

TO.4 Planning the Response

Having robust emergency response plans for schools that are in a tornado-prone region (that is, within the 160, 200, and 250 mph wind zones per Figure TO-7) is particularly important because of the limited warning time that comes with tornadoes. All school management, faculty, staff, students, and parents should respond immediately and know exactly what to do in different situations (e.g., if the warning occurs during class, while on a bus, or during recess). In particular, the following are recommended:

- **Drills.** Drills should be conducted periodically at each school. During drills, students, faculty, and staff should go to the safe room/shelter or designated refuge areas. It is recommended to conduct one drill shortly after the start of the school year and one before spring (an active time for tornadoes). It is also important to provide information on taking refuge when students are not at school (e.g., seek small, windowless interior rooms).

- **Safe Room/Shelter.** For schools that do not have access to a safe room/shelter, plans for adding one should be developed. The case of Plaza Towers Elementary School (Figures TO-12 and TO-13) demonstrates the importance of having access to a safe room/shelter during a tornado. Operations in a safe room during a tornado should be considered in school emergency operations plans. Considerations should include security and safety, first aid, and plans for communications with others outside of the safe room. For more information on this particular topic, readers should see FEMA P-361, which provides emergency management considerations for safe rooms, including communications and emergency supplies.

> If a tornado warning is issued (i.e., a tornado has been spotted), building occupants should move to the tornado safe room(s)/shelter(s), or designated refuge areas if they do not have access to a safe room/shelter. Occupants may be there for several minutes or perhaps an hour or longer, until the warning has been cancelled.
>
> If a tornado strikes the school, debris may preclude family reunification until after first responders remove the debris.

No Safe Room: Plaza Towers Elementary School, Moore, Oklahoma

Plaza Towers Elementary School in Moore, Oklahoma, was constructed in the 1970s with subsequent additions. The one-story school was of common construction of its time (unreinforced concrete masonry unit (CMU) interior and exterior walls, steel joists and steel decking). It was hit by a violent tornado on May 20, 2013 that resulted in the deaths of seven students and injury of many more (NIST, 2013).

The first local tornado warning was issued prior to the normal school dismissal time. About a half hour earlier as the storms approached, all Moore Public School District schools were alerted to move everyone into designated refuge areas. Plaza Towers did not have a safe room/shelter.

Students and teachers were in the "designated area of safety" before the tornado hit. All buildings on the campus sustained significant structural damage (Figure TO-12), including the collapsed refuge area where seven victims were crushed (Figure TO-13). Figure TO-13 shows what remained of the hallway where faculty had been instructed to take their students and where the seven fatalities occurred. Most of the debris had already been cleared at the time the photograph was taken.

The fatalities at this school illustrate the sobering importance of safe rooms/shelters and of conveying a realistic sense of risk to students, faculty, staff, and parents.

Figure TO-12   Post-tornado view. The area circled in red shows the approximate location in the hallway where the 7 fatalities occurred (NIST, 2013; Original photo source: NOAA).

Figure TO-13   Damaged hallway where seven fatalities occurred (most of the debris has already been removed). This hallway area was a "designated area of safety" (NIST, 2013).

- **Best Available Refuge Areas.** As interim measure until a safe room/shelter is made available to the building occupants, a qualified architect or engineer should select best available refuge areas within the school grounds. Until a safe room/shelter is incorporated on school grounds, all school occupants should plan to go to the best available refuge areas during a tornado.

> Typically, advance warning is sufficient for students, faculty, staff and visitors to reach the safe room/shelter or best available refuge areas.

- **Weather Radio.** Schools should be equipped with weather radios, so that office personnel will be aware when the National Weather Service issues tornado watches and warnings.

- **Portable Classrooms.** Portable classrooms are often more susceptible to wind damage than school buildings. It is therefore recommended that portable classrooms not be occupied during time when a tornado watch has been issued by the National Weather Service.

- **Policies and Operational Plans.** Policies and operational plans should include considerations for a tornado watch or warning being issued during different scenarios, such as outdoor events (e.g., sports or ceremonies), school bus operations, and school release. For school bus operations, it is recommended that the policy/plan include the following: (1) buses should not enter an area that is within the boundary of a tornado warning (this will necessitate someone at the bus operations center having access to National Weather Service warnings, and having communications with drivers); and (2) if a bus inadvertently gets close to a tornado, the driver should try to get to find a nearby building and have the bus occupants go inside it (typically it will be better to have students be in even a weak building rather than on a bus). Policies and operational plans should also outline procedures for parents picking up students prior to an event, school lockdown prior to an event, and family reunification after school lockdown.

TO.5 Planning the Recovery

Schools affected by a tornado should be evaluated by an architectural and engineering team to determine whether or not they are safe to reoccupy. ATC-45, *Field Manual: Safety Evaluation of Buildings After Wind Storms and Floods* (ATC, 2004), provides guidance on rating the safety significance of damage. The architectural and engineering team should also determine whether there was damage that is not directly related to life safety. Having pre-event contracts with architectural and engineering teams, and with contractors (including roofing contractors) for immediate post-event damage assessment and emergency repairs is highly recommended.

> In the aftermath of the student fatalities during the Moore 2013 tornado, voters passed a bond measure by a 3 to 1 margin to construct safe rooms at all 23 of the 25 schools within the school district. The other two schools had safe rooms that had been constructed after destructive tornadoes in 1999.

If there is significant damage, an evaluation should be performed to determine if it is cost-effective to repair the damage versus demolish the building and build a new one. If the damage is severe enough, repairs will need to include various code-required upgrades (such as fire alarm systems), which can dramatically increase cost and may economically necessitate replacement with a new building. School leaders should consider incorporating a tornado safe room/shelter as part of the repair or reconstruction if the school is located in the 160, 200, or 250 mph wind speed zone.

TO.6 Recommended Resources

Full citations of all references used to develop this supplement are listed in the References section in this *Guide*. The following is a list of recommended resources that might be useful for school leaders that are addressing school tornado risk. In some cases, a document is both in the References section and listed here as a recommended resource.

ATC-45, *Field Manual: Safety Evaluation of Buildings after Wind Storms and Floods* (ATC, 2004). This document provides guidelines for evaluating whether it is safe to enter and reoccupy buildings damaged by wind or flood. It is intended to be used by designers and building professionals performing safety evaluations. This document does not provide guidance on determining damage and repair costs.

Case Study – School Community Safe Room: Wichita, Kansas (FEMA, 2015a). This document provides a case study of a new middle school classroom wing that was designed to meet or exceed the design criteria in the 2008 editions of ICC 500 and FEMA P-361. https://www.fema.gov/media -library/assets/documents/103902

FEMA P-361, *Safe Rooms for Tornadoes and Hurricanes: Guidance for Community and Residential Safe Rooms* (FEMA, 2015c). This document is primarily intended for architects and engineers; however, it also provides emergency management considerations for the operations and maintenance of tornado safe rooms. www.fema.gov/media-library/assets/documents/3140

FEMA P-424, *Risk Management Series: Design Guide for Improving School Safety in Earthquakes, Floods, and High Winds* (FEMA, 2010a). This document is primarily intended for architects and engineers. It primarily pertains to new schools, but does provide guidance for existing schools. https://www.fema.gov/media-library-data/20130726-1531-20490-0438 /fema424_web.pdf

FEMA P-431, *Tornado Protection: Selecting Refuge Areas in Buildings* (FEMA, 2009b). This document is primarily intended for architects and engineers. www.fema.gov/media-library/assets/documents/2246

ICC 500, *Standard for the Design and Construction of Storm Shelters* (ICC, 2014a). This document sets forth requirements for design and construction of buildings or portions thereof to be used as storm shelters.

JetStream – An Online School for Weather. This website was developed to help educators, emergency managers, or any other interested party in learning about weather and weather safety. http://www.srh.noaa.gov /jetstream/index.html

Mitigation Case Studies: Protecting School Children from Tornadoes, State of Kansas School Shelter Initiative (FEMA, 2002). This document describes the experiences of two Kansas schools affected by the May 3, 1999 tornadoes and the actions taken to mitigate against future events. www.preventionweb .net/files/2755_ksschoolscs1.pdf

Preliminary Reconnaissance of the May 20, 2013, Newcastle-Moore Tornado in Oklahoma (NIST, 2013). This report describes the reconnaissance investigations that were conducted on one critical facility and two educational facilities that were affected by the May 20, 2013 Newcastle-Moore tornado in Oklahoma. http://www.nist.gov/customcf/get_pdf.cfm ?pub_id=914721

Storm Prediction Center. For information on tornado meteorology, see http://www.spc.noaa.gov/faq/tornado/

The StormReady® Program, hosted by the National Weather Service, offers a community recognition program for communities and institutions that implement storm ready activities, including activities specific to preparing for tornadoes. To learn how to become StormReady®, visit http://www .weather.gov/stormready/.

Supplement TS

Tsunamis

All open ocean shorelines in the United States are potentially susceptible to catastrophic flooding by tsunamis, which can destroy facilities and kill people unable to reach safe ground. The risk of tsunami inundation varies considerably. At least 190 schools in California, Oregon, Washington, Alaska, and Hawaii are considered at-risk to tsunami inundation (Saiyed et al., 2017). Estimates of school facilities at risk in other states have not yet been compiled, but geologists, oceanographers, and other experts are working to better understand the tsunami hazard to communities on the islands and coastlines of the Pacific Basin as well as along the U.S. Atlantic Coast and the Gulf of Mexico.

This supplement addresses existing schools and planned new school construction located in and near tsunami hazard zones in the United States and its territories. It provides guidance based on field observations and research conducted on schools and other buildings that have been affected by tsunamis worldwide.

After reading this supplement, school administrators, school emergency managers, teachers, and other school leaders should be able to:

- Determine whether their school is (or might be) located within or near a tsunami hazard zone;
- Understand general considerations and options for existing school buildings in tsunami hazard zones;
- Understand the roles that schools can serve by providing shelter or refuge to evacuated students, staff, and community;
- Create or update a school disaster plan with specific considerations for tsunamis, including evacuation options; and
- Identify aspects that should be considered to facilitate school recovery following a tsunami.

TS.1 Overview of Tsunamis

In the Japanese language, the word "tsunami" means "harbor wave." Often incorrectly referred to as tidal waves, tsunamis have nothing to do with normal tides. The phenomenon is a *series* of waves generated by a powerful physical disturbance within a body of water. In an enclosed body of water, a similar displacement that produces a standing wave is called "seiche" rather than tsunami. When a column of water is displaced (typically by an earthquake, landslide, or volcanic eruption), the waves propagate across the body of water until they reach nearby and distant shores.

In the open ocean, tsunami waves can travel at speeds up to 600 miles per hour. Generally imperceptible in open waters, tsunamis slow down and build in height as they approach the coastline. They flood ashore in a "run-up" that delivers great force and can flood areas far above normal high tide or storm surge levels. Tsunamis with run-up elevations that exceed 3.28 feet (one meter) above normal sea levels are particularly dangerous to people and property, but smaller tsunamis are also life threatening and can cause extensive damage. As little as six inches of fast-moving water can knock over an adult, and children are at risk in even less.

The time elapsed between successive tsunami waves can vary from five to ninety minutes. The first wave is usually neither the largest nor the most significant in a tsunami sequence. The most damaging waves usually occur within the first several hours of the initiating event. Local effects can vary due to many factors, including currents and the near shore underwater environment. One coastal community may experience destructive inundation while another, not far away, may experience no tsunami flooding at all in the same event. These variations cannot always be predicted.

TS.1.1 Causes of Tsunamis

Tsunamis may be caused by undersea earthquakes, landslides occurring either above or below the water surface, volcanic activity, atmospheric activity, and, although rare, meteorite strikes. The most common causes are summarized in Figure TS-1.

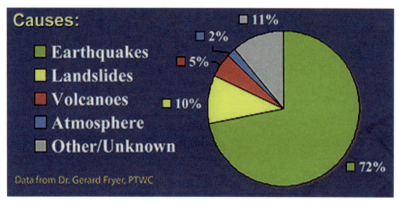

Figure TS-1   Tsunami sources (Pacific Tsunami Warning Center).

Earthquakes account for some 72% of recorded tsunami events. Offshore, underwater earthquakes rapidly displace the seafloor and can shift a column of water upward, generating tsunami waves.

Landslides are the second most common cause of tsunamis, accounting for about 10% of events. Landslides both above and below the water surface have the potential to produce tsunamis even in areas not considered prone to offshore earthquakes. Underwater landslides (for example, down the slope of an undersea river canyon or continental shelf) can essentially 'pull' a column of water downward, causing displacement that generates tsunami waves. Landslides have generated some of the largest tsunamis ever recorded. The 1964 Great Alaska earthquake triggered a landslide responsible for a "mega-tsunami" in Lituya Bay, Alaska with a wave run-up that damaged vegetation as high as 1,740 feet above sea level (Miller, 1960).

About 2% of tsunamis have atmospheric causes. So-called "meteotsunamis" are caused by air pressure disturbances associated with fast moving weather systems such as squall lines (narrow bands of storms and high winds associated with cold fronts). These disturbances can generate ocean waves that travel at the same speed as the overhead weather system. Development of a meteotsunami depends on factors including the intensity, direction, and speed of the disturbance as it travels over a body of water with a depth that enhances wave magnification. These events are more likely to occur in the Gulf of Mexico, along the U.S. East Coast, and in the Great Lakes. A recent meteotsunami struck the Florida coast near Naples during a powerful storm in January 2016.

TS.1.2 Local or Distant?

For the purpose of emergency planning, tsunamis are classified as *local* or *distant*. Local tsunamis originate close to shore, typically from an earthquake along an offshore fault. Because they are usually preceded by a strong earthquake and can reach adjacent shorelines within minutes, local tsunamis are considered the most dangerous.

Natural warning signs typically precede a local tsunami. These signs may include water that recedes from shore in an unusual manner, a water surface that appears to boil or churn, and/or an audible roar. An earthquake of any strength in proximity to a coastal area should always be considered a tsunami warning sign. No official notification may be received prior to the arrival of a local tsunami, which may reach shore within minutes of the earthquake or landslide that causes it. Immediate action, including evacuation to high ground or inland, is essential.

A Tragic Loss in Hawaii

On April 1, 1946, a magnitude-7.1 earthquake off the Aleutian Islands in Alaska caused a destructive tsunami. Locally, the earthquake generated a wave that surged as high as 135 feet, destroying a newly built lighthouse on Unimak Island and killing its crew of five. The tsunami crossed the Pacific Ocean to reach the Hawaiian Islands.

When the tsunami waves reached Laupāhoehoe Point on the north coast of the Big Island of Hawaii, water drew back from shore in an unusual manner and attracted the students and staff of nearby Laupāhoehoe School to tide pools exposed by the drawdown. The waves that followed inundated the shore with a run-up that reached 56 feet above sea level. The waves killed sixteen students and five teachers, destroyed teachers' residences, and flooded the school grounds. A monument commemorates the twenty-four total victims at Laupāhoehoe Point Park.

A Near Miss in Oregon

The Great Alaska earthquake, the largest earthquake recorded in United States history (magnitude-9.2), struck on a fault off Alaska's coast on Good Friday, March 27, 1964. This earthquake caused local devastation including the massive landslide that caused the "mega-tsunami" reported at Lituya Bay, and the tsunami generated by the earthquake itself caused significant damage on the West Coast of the United States and in Hawaii.

In the small community of Cannon Beach, Oregon, approximately 1,500 miles from the earthquake's epicenter, the tsunami lifted a bridge deck spanning Ecola Creek off its abutments and floated it upstream. Parts of the town itself were filled with debris. Cannon Beach Elementary School (the Quonset hut in the top right corner of Figure TS-2), adjacent to Ecola Creek and just a few feet above sea level, somehow escaped destruction. The school continued to operate on the site until 2013.

Figure TS-2   Impacts of the 1964 Alaska tsunami on lower Ecola Creek, Cannon Beach, Oregon. (Photo source: Cannon Beach History Center)

Distant tsunamis are classified as those tsunamis that travel across oceans, causing impacts hundreds to thousands of miles away from their originating source. The long travel distance allows accurate prediction of arrival time.

Although natural warning signs, such as water receding from shore, may precede the arrival of a distant tsunami, official warning systems and announcements from emergency managers can alert affected communities well in advance of any danger, allowing plenty of time for orderly evacuation of facilities and locations at risk.

Either type of tsunami can be dangerous if communities and schools in vulnerable locations are unprepared. Fortunately, with advance preparation (and in some cases mitigation), communities, schools, students, faculty, and staff can expect to survive both local and distant tsunamis.

TS.2   Is Your School in a Tsunami Hazard Zone?

As mentioned in the chapter introduction, all open ocean shorelines in the United States are exposed to tsunamis, but the hazard level varies. Table TS-1 shows the variation in hazard levels, deaths, and damages attributable to tsunami inundation for coastal areas of the United States and its territories since the year 1800. Over roughly two centuries, tsunamis have been responsible for more than 700 deaths and nearly $2 billion in direct damages (in inflation-adjusted dollars) in the United States.

Tsunamis are highly variable, and the task of identifying areas susceptible to flooding by local and distant tsunamis is complex. In general, state geology

Table TS-1   Coastal Areas in the United States Ranked by Tsunami Hazard (NTHMP, 2016)

Location	Tsunami Hazard Level	Number of Events	Deaths Reported	Damage Reported
Hawaii	High to Very High	134	293	$622 million
Alaska	High to Very High	100	222	$688 million
U.S. West Coast	High to Very High	94	25	$241 million
American Samoa	High	68	34	$139 million
Guam & N. Mariana Islands	High	25	1	-
Puerto Rico & U.S. Virgin Islands	High	13	164	$63 million
U.S. Atlantic Coast	Very Low to Low	8	-	-
Alaska Arctic Coast	Very Low	-	-	-
U.S. Gulf Coast	Very Low	1	-	-

Note: "Events" reported since early 19th century. Hazard levels are qualitative and based largely on the historical record through 2014, geological evidence, and location relative to known tsunami sources, all of which provide clues to what might happen in the future. Damage estimates adjusted for inflation to 2016 dollar values.

FEMA P-1000   **TS: Tsunamis Supplement**   **TS-5**

agencies lead or coordinate the task of mapping coastal areas prone to inundation, combining computer simulation models, topographic data, and knowledge gained from tsunamis throughout the world.

Local and state emergency managers consult inundation maps, typically the worst-case scenario displayed, to map tsunami evacuation zones and designate local evacuation routes. Evacuation zones and routes are commonly shared with the general public in the form of evacuation brochures distributed by local emergency managers and public safety agencies. The maps in evacuation brochures depict evacuation zones for the maximum extent of tsunami flooding. Each state makes local evacuation brochures and maps available online as well as in printed form. The National Tsunami Hazard Mitigation Program (NTHMP: a federal-state partnership including the National Oceanic and Atmospheric Administration, the Federal Emergency Management Agency, the U.S. Geological Survey, and 28 U.S. states and territories) has compiled tsunami inundation and evacuation maps, as well as resources specifically useful to schools. This information is available at: http://nws.weather.gov/nthmp/NTHMP_Web_Resources.html.

TS.2.1 Mapped Tsunami Hazard Zones

State geological survey agencies in at least eight coastal states and four offshore territories of the United States have created and published tsunami inundation maps, technical reports, and evacuation brochures.

Evacuation maps typically display inundation zones, areas of higher ground, evacuation routes, and assembly areas. Figure TS-3 shows an example of the evacuation map prepared for Crescent City, a coastal city with approximately 7,500 residents in northern California. On this map, an elementary school, a high school, and a church are indicated as assembly points (gray circles) for people who follow designated evacuation routes to safe ground.

TS.2.2 Unmapped Tsunami Hazard Zones

Technical experts have not assessed all tsunami hazard zones along United States shorelines, but all coastal areas face some degree of tsunami hazard. As natural hazard educators in locations considered lower-risk sometimes say about tsunamis, "They may be rare… Let's still prepare!" Local offshore conditions and on-shore topography make tsunami flooding highly variable from one community to the next. This variability makes local information especially important.

In coastal communities for which no tsunami hazard assessment or evacuation map exists, school leaders can contact local emergency

> For a map of tsunami hazard zones, visit: http://nws.weather.gov/nthmp/NTHMP_Web_Resources.html or contact local or state emergency management agencies.

> A subcommittee of the School Earthquake Safety Initiative recently started compiling a list of schools that are located within tsunami evacuation zones along the Pacific Rim. To see the latest list, visit: www.eeri.org/projects/schools/subcommittees/#tsunami.

management agencies, state geological surveys, or state offices of emergency management for assistance. The NTHMP has developed guidelines for unmapped areas or areas of lower risk. This information is available at http://nws.weather.gov/nthmp/documents/Inundationareaguidelinesforlowhazardareas.pdf

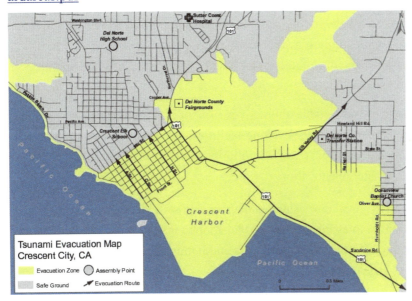

Figure TS-3   Tsunami evacuation map for Crescent City, California. (Photo source: California Geological Survey)

TS.3   Tsunamis and School Buildings

Tsunamis deliver enormous loads to buildings and other structures in their path, and few structures can withstand the forces without sustaining severe damage or complete destruction. The easiest way to eliminate the risk of tsunami damage to schools is to locate school facilities outside of areas subject to tsunami inundation. In every part of the coastal United States, however, schools were constructed in low-lying areas many years before the hazard of tsunami inundation had been recognized. Fortunately, there are steps that all school districts with facilities or property susceptible to tsunamis can take immediately to reduce the risks to their students and staff.

TS.3.1 General Considerations

In an ideal world, all schools would be located on high ground. School districts considering the location and construction of new facilities as part of capital planning efforts can employ a tsunami mitigation strategy of *avoid, minimize, and replace* in order to limit exposure of both the facility and its occupants to tsunami hazards.

> **School Facility Planning Considerations:**
>
> 1. **Avoid** tsunami hazard zones if possible.
> 2. **Minimize** exposure to inundation.
> 3. **Replace** facilities as last resort.

- **Avoid.** In order to improve life safety and reduce the risk of damage and losses to school facilities from tsunamis, avoid siting new or replacement facilities within known or suspected tsunami hazard zones.

- **Minimize.** In communities where facilities cannot be sited outside of known or suspected tsunami hazard zones, risk to life safety may be minimized by integrating *vertical evacuation* into the school facilities located on the highest available ground. Taking advantage of natural elevation will reduce the costs of integrating vertical evacuation into school design.

- **Replace.** If no suitable alternatives exist or if vertical evacuation is not feasible, plan to replace existing facilities, but consider the costs and prepare students, staff, and community for the realities of safe evacuation from a vulnerable location.

School leaders, local emergency managers, and state geological survey staff can collaborate to evaluate proposed locations for school facilities, helping school officials to plan capital budgets and bond or levy measures that include construction funding.

TS.3.2 Existing School Facilities

School districts that have facilities located within tsunami hazard zones should consider crafting and adopting a "risk-reduction hierarchy policy" to prioritize options for the relocation of schools at risk, or the replacement of existing schools with tsunami-resistant structures that incorporate vertical evacuation.

One of the most important priorities for existing school facilities located within known hazard zones is to ensure that practical tsunami evacuation plans exist and are tested by holding regular evacuation drills. Practicing the procedures specified by school or district-wide emergency plans will validate planning assumptions and allow plans to be refined to ensure they will lead to the desired outcome: safe students, teachers, and staff when real-world evacuations occur.

TS.3.3 New Facilities: Design for Vertical Evacuation

Some coastal communities lack natural high ground on which to locate schools. Some lack safe terrain to which students and staff can evacuate in a timely manner in response to a nearby earthquake or a tsunami warning. For these communities, new concepts guiding the integration of artificial "high ground" into the built and natural environment deserve consideration.

Vertical Evacuation Becomes a Reality in Coastal Washington State

Local officials and school leaders dedicated the first engineered vertical evacuation structure in the United States (a Project Safe Haven initiative) in June 2016. Ocosta Elementary School sits at the base of a low, sandy peninsula between Grays Harbor and the Pacific Ocean in southwest Washington State. The $16 million replacement of a 50-year-old deteriorating elementary school includes a tsunami refuge structure designed to accommodate all of the district's 700 students and staff plus up to 1,300 community members at a safe elevation above sea level.

After two prior failures to pass a local bond measure to replace the old school, the new project won at the polls with 70 percent "yes" votes, indicating strong community support for this vision joining education with public safety. The robust engineering and vertical evacuation features of the new school added only $2 million to project costs.

Figure TS-4   Ocosta Elementary School's heavily reinforced vertical evacuation structure under construction in Westport, Washington (Magistrale, 2015).

The concept of *vertical evacuation* as a tsunami risk reduction strategy includes such applications as the design of tsunami-resistant buildings with upper floors or roof used for evacuation, the erection of reinforced earthen berms, the integration of reinforced towers into existing school facilities, and other approaches.

Tsunami vertical evacuation design has advanced in the years since the 2004 Indian Ocean tsunami, as federal guidance and the recommendations of professional engineers evolve. A key consideration facing school districts that seek to incorporate tsunami-resistant design features into their inventory of school buildings is cost. Experience in Washington State, for example, indicates that the incremental cost to include tsunami vertical evacuation in

> **Guidance on Vertical Evacuation**
>
> **FEMA P-646**, *Guidelines for Design of Structures for Vertical Evacuation from Tsunamis* (FEMA, 2012b)
>
> **FEMA P-646A**, *Vertical Evacuation from Tsunamis: A Guide for Community Officials* (FEMA, 2009d)
>
> "Chapter 6, Tsunami Loads and Effects," **ASCE/SEI 7-16**, *Minimum Design Loads and Associated Criteria for Buildings and Other Structures* (ASCE, 2017b)

the construction of new facilities adds 15-20% to overall project budgets (Washington Military Department, 2012).

In the absence of funding sources specifically dedicated to tsunami-resistant construction, school districts should consider other options for funding. Section 5.3 provides general ideas and guidance on funding. In addition, the tsunami program officer in each coastal state's emergency management agency can determine whether funding from the National Tsunami Hazard Mitigation Program may be available to support planning, design, and engineering activities. A vertical evacuation initiative in coastal Washington State (Project Safe Haven) successfully sought funding of this type and used the funds to perform site-specific inundation modeling to determine the height of structures needed to provide safe ground to 6,300 coastal residents.

TS.3.4 Schools as Evacuation Shelters or Refuges

Even if a school is not located within the tsunami hazard zone, it may still be impacted by the tsunami event. Local emergency managers may designate schools *outside* the tsunami hazard zone as emergency refuges, shelters, or assembly points for community evacuation. Fulfilling these functions may have consequences for normal school activities if some or all of a school facility is reassigned to a broader community purpose. There are important logistical and legal considerations when schools are designated as emergency shelters, including the provision of facilities and services that meet accessibility requirements. More on this topic is covered in Section 5.1.5.

TS.4 Planning the Response

> A subcommittee of the School Earthquake Safety Initiative recently compiled a Tsunami Preparedness Checklist for K-12 school administrators and principals. The list can be accessed here: www.eeri.org /projects/schools/subcommittees /#tsunami.

Comprehensive school emergency management plans are essential for schools that face known tsunami hazards, particularly those located in coastal areas where the hazard level is rated "high" to "very high." Schools in tsunami zones must plan for a disaster that is both unpredictable and highly variable in impact, and must take both local and distant tsunami scenarios into account.

Schools in places where a local tsunami would most likely be associated with a strong local earthquake must also plan for self-protective behavior by students and staff during the earthquake and before evacuation from the tsunami hazard, and for the likelihood that earthquake damage may complicate tsunami evacuation routes and procedures. All of the considerations discussed in Section E.4 in the *Earthquakes Supplement* must be taken into account.

TS.4.1 Warning Signs

Tsunami hazards are announced by natural warning signs and by official warning systems. Several natural signs indicate that a tsunami may be imminent:

- **Ground Shaking.** Earthquakes cause most tsunamis. In the event of an earthquake, students and staff should always "Drop, Cover, and Hold On" (even if out-of-doors) for the duration of earth shaking to protect themselves from injury, and should always assume that a tsunami may have been generated by the shaking.

- **Sea Receding.** Retreat of water from the shoreline, resembling an unusually low tide, is a sign that a tsunami may strike.

- **Loud Roar.** An unusually loud roaring sound or other abnormal noise from the sea can indicate that a tsunami is on its way.

One or more of these signs should trigger immediate mandatory evacuation of students and staff to high ground or inland until emergency managers or other public officials provide notice that it is safe to return to low-lying areas.

Two national tsunami warning centers operated by the National Weather Service (NWS) provide coastal communities with advance notice of distant tsunami hazards. The U.S. National Tsunami Warning Center, located in Palmer, Alaska, serves the continental United States and British Columbia. The Pacific Tsunami Warning Center, based in Hawaii, serves Hawaii, Guam, American Samoa, the Commonwealth of the Northern Mariana Islands, the Caribbean, and many island nations throughout the Pacific Ocean.

Both centers use a common hierarchy of alerts to convey the potential tsunami hazard that may exist for various shorelines. The hierarchy includes four alert levels: *Warning*, *Advisory*, *Watch*, and *Information Statement*.

- **Warning.** The highest level of alert, a Warning *requires immediate action*. Upon hearing a Warning, evacuate immediately to higher ground or inland, as a tsunami that could inundate the coast is expected.

- **Advisory.** The second highest level of alert, an Advisory also *requires action*. Flooding on land is not expected, but severe and damaging currents are expected or are already occurring. Protective measures include staying off the beach, staying away from shorelines, and avoiding ports, harbors, and marinas.

- **Watch.** A Watch is generally issued when a strong earthquake has occurred and the Tsunami Warning Centers have not yet determined

> **Official Tsunami Alert Levels**
>
> 1. **Warning.** Requires *immediate* action including evacuation.
>
> 2. **Advisory.** Requires action.
>
> 3. **Watch.** Await further information; possible action.
>
> 4. **Information Statement.** Situation under analysis.

whether a damaging tsunami has been generated. Stay tuned for further information and prepare for the possibility that additional actions, such as evacuation, may be required.

- **Information Statement.** Issued to convey important information that does not require immediate action, for example the fact that a strong earthquake has occurred but no tsunami has been generated. In some instances, a Statement also conveys that the Tsunami Warning Centers are analyzing the situation and additional information or products may be issued.

Warnings and Watches are broadcast over the Emergency Alert System (EAS), a national public warning system. The signal will be relayed over television and radio networks, including All-Hazards Radio (also known as NOAA Weather Radio). Coastal schools should be equipped with small receivers for emergency radio bands, which will alert for tsunami hazards and other potentially dangerous situations including severe weather. Keeping the emergency radio switched "on" in the school office or other area easily accessed by principal and staff when school is in session is a key step in receiving tsunami and severe weather alerts.

School emergency plans should include procedures for each type of alert that may be issued. Administrators, faculty, staff, and students should be familiar with these alert levels and should practice the procedures on a routine basis. Because time may be of the essence during an actual tsunami emergency, pre-scripting emergency messages that can be read or broadcast over school public address and notification systems can save time.

TS.4.2 Plans, Policies, Procedures

Coastal schools need emergency plans that address the full range of hazards from local and distant tsunamis. Plans should include a variety of scenarios of response.

School leaders need to clearly understand whether their schools are susceptible to local tsunamis, distant tsunamis, or both. A safety plan must address the particulars of each situation. For example, schools within the small city of Hoquiam, Washington, located on Grays Harbor near the Pacific Ocean, are susceptible to local tsunamis from nearby sources including earthquakes on the Cascadia Subduction Zone, and to distant tsunamis that can originate in other areas around the Pacific Rim. The Hoquiam School District has adopted tsunami safety procedures, summarized on the next page, that are time-dependent and based on a threshold deemed appropriate to each situation.

Example of Tsunami Safety Procedures: Hoquiam School District

Following consultation with the Grays Harbor Office of Emergency Management, the Hoquiam School District will observe the following procedures in the event of a Tsunami:

Situation 1

There is notification to the Superintendent's office of a Tsunami impacting the community in five hours or less.

Our Response:

1. Students and staff will immediately evacuate to designated "high ground" location.

- Emerson, Middle School and High School to the High School baseball field.

- Central and Lincoln went to the City Cemetery.

2. Phone message will be made to parents/guardians through the Superintendent's Office (if possible).

3. Schools will remain at high ground until notified by the Superintendent/designee.

Situation 2

There is notification to the Superintendent's office of a Tsunami impacting the community in more than five hours.

Our Response:

1. Phone message will be made through the Superintendent's Office to parents/guardians.

2. Superintendent will notify media.

3. Parents/guardians will be permitted to pick up child from the individual school up to 2 hours before impact or prior to the end of the school day.

4. Two hours before the anticipated Tsunami or at the end of the school day all remaining students will be bussed to the high school to gather in designated classrooms on the upper campus.

5. The upper high school staff parking lot will be designated as the pick-up zone for parents/guardians wishing to pick up their student.

6. A staff member(s) will be assigned to meet parent/guardian in the upper HS staff parking lot.

7. Designated staff member(s) will retrieve the student(s) for the parent/guardian after verifying parent/guardian identity.

8. Staff and students will remain on the campus until notified by emergency management officials.

Based on your facility's layout and location, more refined procedures may be necessary. It may also be prudent for students, faculty, and staff to understand the appropriate response and protocol if they are outdoors for physical education as opposed to inside a classroom. Beyond discussing this information, or posting it in conspicuous locations in each classroom can help reinforce the desired action.

Hoquiam's emergency plan specifies the high-ground locations to which students and school personnel will evacuate in the event of a local tsunami. These refuge and assembly locations are communicated to parents and guardians so that, in the event of an actual tsunami response, parents anxious to pick up their children will not rush into a potentially dangerous situation at the school itself.

In areas where tsunami hazard zones have been mapped, emergency officials typically designate official evacuation routes and assembly areas. School leaders should walk the official evacuation routes themselves, and join evacuation drills with school staff and students. Routes for evacuation on foot and routes intended for evacuation by car may differ. Communities that face a high risk of local tsunamis will likely have routes designated for pedestrians only, as a strong local earthquake would be expected to damage or block roadways, making them unsafe for evacuations by car or school bus.

Reviewing the plans regularly and practicing associated procedures is critical to ensuring school safety. By training school staff and students and by conducting routine tsunami evacuation drills, school leaders will develop confidence that their plans can be successfully implemented when needed.

Only by conducting comprehensive drills is it possible to learn how long it may take to evacuate an entire student body to higher ground, following designated evacuation routes. If an evacuation drill takes longer than the interval of time anticipated before a local tsunami arrives, school leaders will know they have to revise plans or seriously consider other possibilities, such as options for vertical evacuation.

School safety drills should be a family affair whenever possible. Inviting parents and guardians to participate in tsunami evacuation drills will allow them to experience an orderly evacuation and equip them to talk with their child(ren) about tsunami hazards and the importance of emergency preparedness. This may prompt improved family emergency preparedness at home. It also gives school personnel and parents and guardians the opportunity to practice post-emergency reunification procedures and to identify opportunities for improvement.

Participation in organized statewide exercises like the annual Great ShakeOut Earthquake Drills held in many states can offer coastal communities an opportunity to link "Drop, Cover, and Hold On" earthquake safety practice with a tsunami evacuation component tailored to local plans and conditions. The TsunamiZone program, with a similar focus on tsunami preparedness activities, is attracting participation in a growing number of states and territories.

The National Weather Service TsunamiReady® recognition program has established guidelines pertaining to tsunami hazard mitigation, preparedness, and response. Participating communities are eligible for benefits including technical assistance and points from FEMA's Community Rating System, which can result in discounts on federal flood insurance premiums. While most participants are counties, municipalities, and military bases, school districts participating in the TsunamiReady® program include Los Angeles Unified School District in California and Lincoln County School District in Oregon.

Preparation for the immediate evacuation that a local tsunami may require includes assembling basic emergency supplies for students and staff, as reunification with parents may prove challenging or impossible due to damage and debris. Go-Kits, which typically consist of a portable container like a backpack or bucket filled with emergency supplies, are especially important in areas susceptible to local tsunamis. These kits provide essential supplies for safety and sustenance until emergency assistance can arrive. Both school administrators and classroom teachers should be supplied with appropriate Go-Kits.

The Lincoln County School District on the central Oregon coast serves a high proportion of students in poverty. Though no district schools are located in the tsunami zone, the communities served face earthquake and tsunami hazards. To ensure that essential emergency supplies are available to all of its 5,200 students in the event of a natural disaster, the district has established disaster caches in shipping containers located at or near every school site. Each cache is stocked with supplies sufficient to meet the basic needs of students and staff for days or weeks following a worst-case Cascadia Subduction Zone earthquake and accompanying local tsunami. Developed with public- and private-sector local partners, the disaster caches are equipped to serve local community colleges and private schools as well as the public school population.

TS.4.3 Evacuation Protocols and Practice

"Practice Makes Perfect." Nothing could ring more true when faced with the immediacy of an impending tsunami, especially in a school or classroom setting. It is only with advance preparation, education, drills, training, and by adopting a tsunami risk reduction hierarchy, that school districts can ensure life safety during tsunami emergencies. Occupants of school facilities need

Tsunami Evacuation Essentials
1. Know where to **GO**.
2. Know what to **TAKE**.
3. Know what to **DO**.

to know where to go, what to take, and what to do. Figure TS-5 illustrates a typical tsunami evacuation sign. All must respond without hesitation, because there may be only one chance to get it right.

Figure TS-5   Tsunami evacuation route sign. (Photo source: James Sherrett)

As noted above in the Hoquiam example on page TS-13, two types of tsunami evacuation protocols should be considered by the school administration: one for local tsunamis and one for distant tsunamis. In the event of a local earthquake, evacuation of buildings and school grounds located within a tsunami hazard zone should occur immediately following earthquake shaking; no additional information is needed.

School leaders should evacuate students from classrooms and grounds to the nearest available natural high ground or designated vertical evacuation structure as soon as ground shaking stops and be prepared to respond to changing conditions if the first assembly location appears threatened. Students and staff should remain in assembly locations until local emergency managers or other officials advise that it is safe to return. All-hazards weather radio will provide current information.

Unlike many other types of emergencies, the tsunami risk can last for up to 24 hours, so evacuation may be enforced for at least that long and possibly longer. All students and staff should stay clear of the tsunami hazard zone until local officials have declared it safe to return. The cancellation of a warning or advisory issued by a tsunami warning center does not mean it is safe to return, because waves and currents are influenced by local conditions that the national warning centers cannot take into account.

Designated evacuation assembly location(s) must be clearly communicated in advance to parents and caregivers, so that they know not to rush to the school itself to pick up their children. Failure to communicate locations for reunification in advance could send parents and guardians into harm's way.

Working in partnership with local emergency managers, school administrators should determine, based on clearance times, whether family reunification can take place at the school itself, or whether reunification must occur at the evacuation assembly location(s) on high ground.

Schools that include a broad age range may consider partnering younger students with older students as "tsunami buddies." High school and middle school students may be able to help elementary students evacuate safely and keep younger students together, maintaining calm and easing anxieties.

Distant tsunamis, which originate from distant sources and require two or more hours to reach an affected shore, change the consideration of evacuation options. Timing must be taken into account: both the time of day

---

Lessons on Tsunami Evacuation that can Save Lives

On March 11, 2011, a magnitude 9.0 earthquake struck off the island of Honshu, Japan, generating a tsunami that killed more than 15,800 people.

The Kamaishi Higashi Middle School and Unosumai Elementary School in Kamaishi City were located within a mapped tsunami hazard zone. To improve student preparedness for tsunami evacuation, students in these two schools had routinely practiced evacuation drills since 2005. Their teachers devoted five to ten hours of class time each year to tsunami hazards.

When the earthquake struck but before the tsunami waves arrived, students first planned to evacuate to the third floor of their building. However, observing a tsunami higher than expected, the students abandoned those plans and relocated to higher ground away from the school. They relocated twice more during the event because of the extent of inundation. While more than 1,000 people lost their lives in Kamaishi City, none of the casualties were school-age children. Classroom training had emphasized that students should assess the situation as they see it and be able to respond to changing events. This training provides the best explanation for the survival rate that has become known as the "Miracle of Kamaishi."

On the other hand, in communities of Northern Honshu including Taro, Kesennuma, and Ishinomaki, while students were evacuating to safety, many parents who rushed directly into the tsunami hazard zone to retrieve their children tragically lost their lives. These avoidable fatalities underscore that children and parents alike need to know the details of school preparedness and evacuation plans.

FEMA P-1000   TS: Tsunamis Supplement   TS-17

Lessons on Tsunami Evacuation that can Save Lives (continued)

Togura Elementary School located in the coastal part of Minami-Sanriku in Miyagi Prefecture, shown in Figure TS-6, was struck by the tsunami. Inundation submerged the three-story school up to its roofline, and the building itself suffered severe damage as the waters receded, with most nearby buildings being washed away entirely.

Staff and students assembled in the schoolyard immediately following the earthquake, and then evacuated to a designated safe site on high ground nearby. The tsunami then bore down upon them, but everyone was able to escape unharmed—due to a quick reappraisal in the midst of the disaster.

Because this elementary school was in a location directly exposed to tsunami hazard, the school's evacuation drills took a range of scenarios into consideration. Thanks to their routine discussions of tsunami risk, the teachers knew to keep their attention on the ocean even after they had made a secondary evacuation to higher ground. Their ability to recognize the warning signs of a tsunami exceeding all prior experience led them to guide the children to a shrine on even higher ground, saving lives.

Togura Elementary School was rebuilt as part of Minami-Sanriku's municipal recovery efforts and reopened for classes on August 31, 2015.

Figure TS-6   Waves surge past Togura Elementary School in Minami-Sanriku, Japan during the 2011 Great East Japan Tsunami. (Photo source: Ichiro Abe, *The Asahi Shimbun*)

that the tsunami is expected to arrive and the time of day that notification is received will influence the options available to school decision makers.

For example, if an earthquake in Japan causes a tsunami expected to reach the U.S. West Coast in 14 hours and the warning is issued at the end of the school day, school leaders may take no immediate action other than to monitor the event, communicate the status to parents, faculty, and staff, and prepare for possible school cancellation the next morning based on updated information. However, if the school day has just begun and a tsunami warning or advisory indicates arrival within four hours, decisions may be different. Emergency plans and evacuation protocols should anticipate a range of possible scenarios for distant tsunamis.

TS.5   Planning the Recovery

After a tsunami, local officials will begin to assess damage to the community and critical facilities and will inform school administrators if and when it is safe to reoccupy schools. While the tsunami event itself may last up to 24 hours, it may cause other dangers that delay return to the site even longer. Depending upon the amount of damage to the area and hazardous debris left behind, it may be days, weeks, or even months before it is possible to return to areas that were inundated.

Schools that have sustained damage in a tsunami will require inspection by a qualified professional (architect, engineer, or other building professional) to determine whether the structure can be reoccupied. Schools that cannot be immediately reopened for classes should begin implementing their plans for educational continuity (see Section 5.1.4). These may include options such as school sharing or relocating classes and activities into temporary facilities in the near term, and rebuilding schools in a new location outside of the hazard zone in the longer term.

The experience of evacuation from a tsunami can be traumatic for children and adults alike. Even students and staff who have safely evacuated from school may experience other losses in their personal lives. Resuming normal activities including school routines may prove difficult. Schools are in a position to deliver psychological first aid and to support recovery. More guidance on this topic is provided in Section 5.2.

TS.6   Recommended Resources

Full citations of all references used to develop this supplement are listed in the References section in this *Guide*. The following is a list of recommended resources that might be useful for school leaders that are addressing school tsunami risk. In some cases, a document is both in the References section and listed here as a recommended resource.

General Information and Key Programs. The National Tsunami Hazard Mitigation Program (NTHMP) is a coordinated national effort joining the National Oceanic and Atmospheric Administration (NOAA), Federal

Emergency Management Agency (FEMA), the U.S. Geological Survey (USGS), and 28 U.S. states and territories. The purpose is to assess the tsunami hazard, prepare community response, issue timely and effective warnings, and mitigate damage. Key websites of the program and its partners include the following:

- FEMA: http://www.fema.gov and https://www.ready.gov/tsunamis

- National Tsunami Warning Center: http://ntwc.arh.noaa.gov

- National Weather Service: http://www.tsunami.gov

- NTHMP: http://nws.weather.gov/nthmp/

- NWS Tsunami Safety: http://www.nws.noaa.gov/om/Tsunami/

- Pacific Tsunami Warning Center: http://ptwc.weather.gov

- USGS: http://water.usgs.gov/edu/tsunamishazards.html

Maps of Tsunami Hazard Zones. The NTHMP maintains a portal to tsunami hazard zone maps created by the program's participating states and territories. Most state geology agencies and emergency management agencies are also a good source for mapping of coastal hazards. For the NTHMP Map portal, visit: http://nws.weather.gov/nthmp/maps.html.

For guidance for unmapped or low-risk coastal areas, visit: http://nws .weather.gov/nthmp/documents/Inundationareaguidelinesforlowhazardareas .pdf

Drills and Public Engagement Activities. Examples nationwide programs and campaigns are as follows:

- The Great ShakeOut earthquake safety drills have engaged more than 20 million people in all parts of the United States in basic safety practice. To participate, visit http://www.shakeout.org/

- Tsunami Preparedness Campaigns are supported by the National Tsunami Hazard Mitigation Program. To learn more about dates and activities in your area, visit http://nws.weather.gov/nthmp/tpw/tsunami -preparedness-week.html

- The TsunamiReady® program, hosted by the National Weather Service, offers a community recognition program for communities and institutions that implement tsunami preparedness activities. To learn how to become TsunamiReady®, visit http://www.tsunamiready.noaa.gov

- The TsunamiZone is a newer initiative based on a similar awareness-promotion model, promoting tsunami safety and preparedness activities along vulnerable coastlines. To participate, visit http://www.tsunami zone.org/

Vertical Evacuation Options. Vertical evacuation is an emerging concept in coastal facilities design. The following sources may help school leaders begin to evaluate the option for school facilities:

- The American Society of Civil Engineers publishes technical guidance incorporated into building codes; up-to-date information pertaining to tsunamis can be found in "Chapter 6, Tsunami Loads and Effects," in the 2016 edition of ASCE/SEI 7, *Minimum Design Loads and Associated Criteria for Buildings and Other Structures* (ASCE, 2017b).

- FEMA P-646, *Guidelines for Design of Structures for Vertical Evacuation from Tsunamis, Second Edition* (FEMA, 2012b), published in April 2012, is available as a PDF at this link http://www.fema.gov /media-library-data/1426211456953-f02dffee4679d659f62f414639af a806/FEMAP-646_508.pdf

- FEMA P-646 is discussed in a 5-minute video prepared by the Cascadia Region Earthquake Workgroup in early 2012 https://www.youtube.com / watch?v=_h26_DUKMzA

- FEMA P-646A, *Vertical Evacuation from Tsunamis: A Guide for Community Officials* (FEMA, 2009d), published in 2009, is available as a PDF at this link: https://www.fema.gov/media-library-data/20130726 -1719-25045-1822/fema_p646a.pdf

- Project Safe Haven video. A 12-minute video about the integration of vertical evacuation into the design of Ocosta Elementary School in coastal Washington State, a Project Safe Haven initiative and the first tsunami vertical evacuation building constructed in the United States, is available here https://www.youtube.com/watch?v=otI7bUrUOmI

- Washington State's "Project Safe Haven" has explored a variety of vertical evacuation applications including natural berms and evacuation structures. http://mil.wa.gov/uploads/pdf/emergency-management/haz _safehaven_report_pacific.pdf

Tsunami Science. Many print and online sources offer a thorough introduction to tsunami science. A good place to begin is with the U.S. Geological Survey:

- "Life of a Tsunami." http://walrus.wr.usgs.gov/tsunami/basics.html

- USGS Tsunami and Earthquake Research. http://walrus.wr.usgs .gov/tsunami/

Supplement W

High Winds

Most wind damage is caused by tornadoes and hurricanes. However, damage is occasionally caused by other high winds, notably straight-line and down-slope winds. This supplement is applicable to all schools outside of hurricane-prone regions.

This supplement provides guidance for existing buildings, and guidance for new schools that are in the planning stage. After reading this supplement, school administrators, school emergency managers, teachers, and other school leaders should be able to:

- Identify opportunities for incorporating special high wind design enhancements in new facilities to achieve greater resilience;

- Understand the importance of having yearly inspections by maintenance personnel;

- Create or update a school disaster plan with specific considerations for high winds; and

- Identify aspects that should be considered to facilitate school recovery following a high wind event.

W.1 Overview of High Winds

Straight-line winds are the most common and generally blow in a straight line, as opposed to tornado and hurricane winds, which consist of circular winds. They occur throughout the United States and its territories. Down-slope winds blow down the slope of mountains and frequently occur with very high speeds in Alaska, Colorado, and Utah. In the continental United States, mountainous areas are referred to as "special wind regions."

Figure W-1 shows damage caused by straight-line winds. When straight-line or down-slope wind causes school damage, typically only one school in the community is damaged—unlike hurricanes that can damage many or most of the schools. This is typically the case because straight-line or down-slope wind damage usually occurs when a school has a weakness in its building envelope (e.g., roof covering, wall coverings, and windows). These weaknesses are typically caused by inadequate design, workmanship deterioration, or by material degradation. Normally, only one school in the community will have sufficient weakness to be damaged during a given straight-line or down-slope wind event.

In communities with only one school, damage can adversely affect the community; particularly those located in remote areas.

Figure W-1   The roof on this new school blew off during moderate winds soon after it was installed. It failed even though it was a robust system because of a major installation error. (Photo source: Thomas Smith)

W.2   Is Your School in a Region Exposed to High Winds?

Straight-line winds that have the potential to cause building damage can occur anywhere in the United States and its territories. If a school is outside of the "high wind and heavy rain" (red) area of the hurricane hazard map shown in Figure 2-2, damaging winds can still affect the area and this supplement is applicable. Down-slope winds occur in mountainous areas.

W.3   Making Buildings Safer

This section provides guidance pertaining to making buildings more resilient to high wind events.

W.3.1   Existing School Buildings

Typically, unless there is an issue that causes concern, wind vulnerability assessments are not performed on buildings outside of hurricane-prone regions. However, the exterior of the building should be inspected yearly by maintenance personnel. Items that have deteriorated (such as leaky roofs) or become loose (such as rooftop equipment) should be repaired or replaced before they are damaged by wind.

W.3.2 New School Buildings

The provisions in the *International Building Code* (ICC, 2014b) are generally sufficient to provide satisfactory performance when schools experience straight-line or down-slope winds. However, FEMA P-424, *Risk Management Series: Design Guide for Improving School Safety in Earthquakes, Floods, and High Winds* (FEMA, 2010a), provides recommended enhancements and best practices to further avoid damage and occupancy disruptions. The recommendations for schools exposed to straight-line and down-slope winds are much less stringent than the recommendations for schools in hurricane- or tornado-prone regions. Hence, the cost for implementing the recommendations for schools exposed to straight-line and down-slope winds are typically relatively insignificant.

For schools located in remote areas or where the design wind speed is greater than 120 miles per hour, FEMA P-424 recommends the incorporation of additional design, construction, and maintenance enhancements, and provides enhancement recommendations. To determine the design wind speed for a school, go to http://windspeed.atcouncil.org/ and enter the school's address. The output shows design speeds for different risk categories; schools are predominately Risk Category III.

In some areas (such as Alaska), damaging straight-line and down-slope winds often occur during winter months (Figure W-2). Damage repairs during cold weather can be very costly. Also, if the school is in a remote area, repair materials and contractors may need to be brought in by airplane, thus greatly increasing the costs.

Figure W-2   The roof on this school blew off during cold weather, thus increasing the repair costs. (Photo source: Thomas Smith)

W.4 Planning the Response

If high winds are forecast, objects that could be damaged or moved by winds should be secured or moved. All windows should be closed to keep out rain, dust, and debris. When winds above 60 miles per hour are forecast, potentially hazardous areas should be evacuated and avoided. This includes areas adjacent to weak non-load-bearing walls, weak glass curtain walls, and areas below weak long-span roof structures (e.g., auditoriums, gymnasiums). These building elements should be assumed to be weak, unless a wind vulnerability assessment by an architect or engineer has determined that these areas can be safely occupied during high wind conditions. Portable classrooms are often more susceptible to wind damage than school buildings. It is therefore recommended that portable classrooms not be occupied when high winds are forecast.

School leaders and management should develop policies and operational plans regarding what to do if high winds are forecast during various situations, including outdoor athletic events or school bus operations. Plans should consider the following:

- Activities may need to be moved indoors.

- High winds can blow down trees, utility lines and poles.

- Electricity and phone outages may occur.

- Traffic patterns and walking routes can be affected due to debris and power outages.

FEMA P-424 provides additional recommendations to minimize the risk of injury and death during a high wind event.

W.5 Planning the Recovery

Schools that experience wind damage should be evaluated by an architectural and engineering team to determine whether or not they are safe to reoccupy. ATC-45, *Field Manual: Safety Evaluation of Buildings After Wind Storms and Floods* (ATC, 2004), provides guidance on rating the safety significance of damage. The architectural and engineering team should also determine whether or not there was damage that is not directly related to life safety.

If there is significant damage, an evaluation should be performed to determine if it is cost-effective to repair the damage versus demolish the building and build a new one. If the damage is severe enough, repairs will need to include various code-required upgrades (such as fire alarm systems), which can dramatically increase cost and may economically necessitate replacement with a new building.

If new construction is necessary, school leaders should consider implementing the guidance provided in Section W.3.2. If the damage is repairable, in addition to performing repairs, it is recommended to incorporate the enhancements provided in FEMA P-424.

W.6   Recommended Resources

Full citations of all references used to develop this supplement are listed in the References section in this *Guide*. The following is a list of recommended resources that might be useful for school leaders that are addressing the risk from high winds in schools. In some cases, a document is both in the References section and listed here as a recommended resource.

ATC-45, *Field Manual: Safety Evaluation of Buildings after Wind Storms and Floods* (ATC, 2004). This document provides guidelines for evaluating whether it is safe to enter and reoccupy buildings damaged by wind or flood. It is intended to be used by designers and building professionals performing safety evaluations. This document does not provide guidance on determining damage and repair costs.

JetStream – An Online School for Weather. This website was developed to help educators, emergency managers, or any other interested party in learning about weather and weather safety. http://www.srh.noaa.gov /jetstream/index.html

FEMA P-424, *Design Guide for Improving School Safety in Earthquakes, Floods, and High Winds* (FEMA, 2010a). This document is primarily intended for architects and engineers. It primarily pertains to new schools, but does provide guidance for existing schools. http://www.fema.gov/media -library/assets/documents/5264

The StormReady® Program, hosted by the National Weather Service, offers a community recognition program for communities and institutions that implement storm ready activities. To learn how to become StormReady®, visit http://www.weather.gov/stormready/.

Supplement X

Other Hazards

This supplement will briefly describe additional common hazards and is not intended to provide complete guidance, but only to provide general information and point toward additional resources. Hazards discussed in this supplement include:

- Snow storms;
- Volcanic eruptions; and
- Wildfires.

This supplement does not provide an exhaustive list of other natural hazards, such as drought and lighting. Schools located in areas that are vulnerable to other natural hazards should seek other existing guidance on these natural hazards.

X.1   Snow Storms

Winter snow storms and extreme cold weather can have a significant effect on school buildings and their occupants. It is important to monitor the National Weather Service for winter storm alerts and always maintain a functioning NOAA Weather Radio at school and be prepared to respond to alerts and warnings accordingly.

School roof collapse due to accumulation of snow is relatively rare. The cause of collapse can include design defects, construction defects, and deterioration of the roof structure due to roof leaks; however, most collapses are due to snow drifts. Modern snow drift provisions were first introduced in 1988, but many local building codes did not incorporate them until years later. Accordingly, some schools designed in or prior to the 1990s may have inadequate structural capacity to resist drifts. Roof collapse from snow load can also be caused by the addition of new roof insulation which could cause more snow to accumulate because of the reduction in snow melt.

> Some schools designed in or before the 1990s may be susceptible to roof collapse from drifting snow.

FEMA P-957, *Snow Load Safety Guide* (FEMA, 2013c), provides information on identifying potentially vulnerable roof framing systems, and provides a general methodology to monitor buildings for signs of potential failure so that steps can be taken to reduce the potential risk of snow-load-induced structural failure.

Sometimes it is prudent to have snow removed from roofs. However, snow removal involves risk to workers and can cause roof damage. FEMA P-957 recommends that a licensed design professional be retained to determine when removal is prudent. It also recommends that snow removal be performed by a licensed, insured professional roofing contractor who has experience in removing snow from roofs. More specific guidance on snow removal can be found in FEMA P-957.

X.2 Volcanic Eruptions

Schools located in states with active volcanoes (Alaska, California, Oregon Hawaii, Idaho, Washington, and Wyoming) may need to consider one or more volcanic hazards within their emergency planning, preparedness, and response efforts.

Each volcano is different. The potential impacts to school facilities will vary depending upon the distance from the school to the volcano. Hazards associated with volcanoes include lahars (large mudflows), lava flows, and volcanic ash. In 2015, volcanic lava flow from Kilauea volcano in Hawaii disrupted regular school routines and forced the shutdown of schools in the town of Pahoa.

Volcanic ash deposits of as little as $^1/_8$-inch can result in clogged roof drains that can affect roof framing and drainage systems during heavy rainfall. Additionally, ash accumulations on roof membranes become heavy additional loads, which can affect the stability of the roof framing. Removal of volcanic ash is a proactive measure that should be part of a school hazard safety plan.

Volcanoes provide advanced warning of eruptions, which should allow school officials to take proactive measures well before potential issues occur. However, in rare cases, volcanic hazards such as lahars can occur even in the absence of pre-eruption signs. Check with your state geological survey or regional U.S. Geological Survey (USGS) volcano observatory to learn if this is a potential hazard at your site.

The USGS, through its volcano observatories, constantly monitors active volcanoes in the United States, its commonwealths, and territories for any signs of an eruption. USGS has developed a web-based location for information on volcano hazards and their impacts (see: http://volcanoes .usgs.gov/ vhp/ hazards.html).

X.3   Wildfires

A school's susceptibility to wildfire damage is primarily related to the availability of fire-suppression personnel and the following:

- **Defensible Space.** This refers to an area where combustible material, including vegetation, has been treated, cleared, or modified to slow the rate and intensity of an advancing wildfire. Schools surrounded by zones of non-vegetated areas or areas populated by fire-resistant vegetation are more likely to survive.

- **Exterior Building Envelope.** Combustible exterior building components such as roof coverings and exterior walls can be ignited by wildfire, leading to severe damage to or total loss of the school.

Wildfire frequently leads to mudslides and floods. FEMA P-737, *Home Builder's Guide to Construction in Wildfire Zones* (FEMA, 2008), provides information on wildfire hazard and risk assessment, and recommendations for design and construction of new buildings. For existing buildings, it recommends a vulnerability assessment and development of a customized, prioritized list of recommendations for remedial work on defensible space and the building envelope. Although FEMA P-737 pertains to residences, much of the guidance is also applicable to schools.

X.4   Recommended Resources

Full citations of all references used to develop this supplement are listed in the References section in this *Guide*. The following is a list of recommended resources that might be useful for school leaders that are addressing school risk from snow storms, volcanic eruptions, or wildfires. In some cases, a document is both in the References section and listed here as a recommended resource.

Snow Storms. Recommended resources for schools in regions where snow storms are a hazard include:

FEMA P-957, *Snow Load Safety Guide* (FEMA, 2013c) provides information on identifying potentially vulnerable roof framing systems, and provides a general methodology to monitor buildings for signs of potential failure so that steps can be taken to reduce the potential risk of snow-load-induced structural failure. To access the report, visit: https://www.fema.gov/media-library/assets/documents/83501.

For a brief overview of FEMA P-957, visit: https://www.fema.gov/media-library-data/1392984631969-ac57339deb6ee839a52b16b01eeee53e/FEMA_Snow_Load_508.pdf.

Volcanic Eruptions. The USGS provides information on volcano hazards and their impacts. For more information, visit: http://volcanoes.usgs.gov /vhp/ hazards.html.

Wildfires. Resources for schools in regions where wildfires are a hazard include:

FEMA P-737, *Home Builder's Guide to Construction in Wildfire Zones* (FEMA, 2008) provides information on wildfire hazard and risk assessment, and recommendations for design and construction of new buildings. For existing buildings, it recommends a vulnerability assessment and development of a customized, prioritized list of recommendations for remedial work on defensible space and the building envelope. Although FEMA P-737 pertains to residences, much of the guidance is also applicable to schools. To access the report, visit: https://www.fema.gov/media-library /assets/documents/15962.

Prepare Your People for Wildfire Safety: K-12 Schools (PrepareAthon, 2014). This document provides general guidance on preparing for wildfires for K-12 schools. To access this document, visit: https://dphhs.mt.gov /Portals/85/publichealth/documents/School%20Health/Safe%20School %20Environment/Natural%20Disasters/FEMA%20Prepare%20for %20Wildfires%20K-12%20Schools.pdf

Understanding Fire Danger website: https://www.nps.gov/fire/ wildland -fire/learning-center/fire-in-depth/understanding-fire-danger.cfm

Wildfires and Schools (National Clearinghouse for Educational Facilities, 2008b). This document provides guidance that was adapted for schools from FEMA's Wildfires homepage and Wildfire Mitigation Fact Sheet, *Rebuilding After a Fire*. To access this document, visit: http://www.ncef.org/pubs /wildfires.pdf

Wildfire Safety: Social Media Content Shareables. This document provides a list of resources and tips for wildfire safety. To access the document, visit: https://www.fema.gov/media-library-data/1438198310297-b58b3f3d40fc7ae 964927c0bbefdd35f/Wildfire_Safety_Social_Toolkit_2015_Final.pdf

Appendix E

Earthquake Appendix

The purpose of this appendix is to provide more detailed instructions on how to determine the level of seismicity of a specific site and to provide more detailed information on the adequacy of building codes depending on the building type.

AE.1 Determination of Seismicity Region of Site

Because the scale of Figure E-3 is small, it might be difficult to determine a location's seismicity region from the map. This section provides instructions on how to accurately determine the appropriate region.

The United States Geological Survey, which produces the seismic maps used in U.S. building codes, has a calculator tool that will determine seismic design parameters that can be used to categorize a site into the seismicity regions shown in Figure E-3. The steps needed to make such a determination are as follows:

1. Go to the site: http://earthquake.usgs.gov/designmaps/us/application.php. The screen is shown in Figure AE-1.

2. Enter the site address in the box indicated.

3. Select "2012 IBC" from the drop-down menu at location 3.

4. Select "Site Class B – Rock" from the drop-down menu at location 4. If more is known about soils at the site, Class C or Class D might be selected. Many sites are Class D.

5. Select "II, III, or IV" from the drop-down menu at location 5 for Risk Category. For schools with less than 250 students, select II. For typical schools, select III. For schools used as shelters, select IV.

6. Click on "Compute Values" at location 6.

7. The screen shown in Figure AE-2 will come up.

8. Using the values for "S_s" and "S_1" from the box at location 8, go to Table AE-1.

9. Table AE-1 can then be used to determine the appropriate seismicity region, assuming the highest seismicity level determined from the parameters in Table AE-1 governs.

FEMA P-1000 · AE: Earthquake Appendix · AE-1

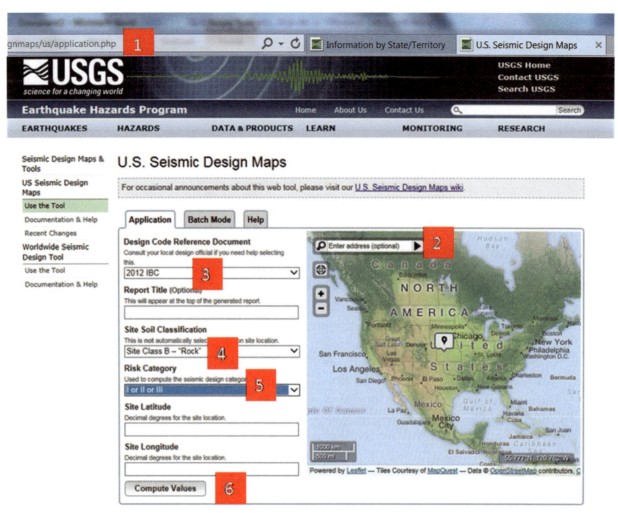

Figure AE-1　　Screenshot from USGS seismic design parameter calculator.

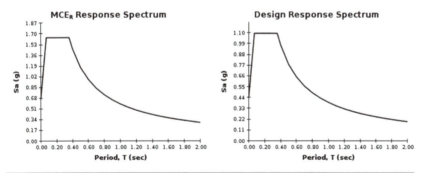

Figure AE-2   Screenshot showing parameters to determine Seismicity Region.

Table AE-1 Determination of Seismic Region from USGS Design Parameters, S_s and S_1

Seismicity Region		Spectral Acceleration Response, S_S (short-period, or 0.2 seconds)	Spectral Acceleration Response, S_1 (long-period, or 1.0 second)
	Low	less than 0.250g	less than 0.100g
	Moderate	greater than or equal to 0.250g but less than 0.500g	greater than or equal to 0.100g but less than 0.200g
	Moderately High	greater than or equal to 0.500g but less than 1.000g	greater than or equal to 0.200g but less than 0.400g
	High	greater than or equal to 1.000g but less than 1.500g	greater than or equal to 0.400g but less than 0.600g
	Very High	greater than or equal to 1.500g	greater than or equal to 0.600g

Note: g = acceleration of gravity in horizontal direction

AE.2 Adequacy of Building Codes

Seismic provisions in building codes have been in place in some regions since 1925. The codes have evolved considerably since that time. Seismic provisions were adopted by different jurisdictions at different times. Buildings built in accordance with the 1976 *Uniform Building Code* (or later) or the *2000 International Building Code* (or later) are often considered to have a low probability of collapse, even in a major earthquake. FEMA P-154, *Rapid Visual Screening of Buildings for Potential Seismic Hazards* (FEMA, 2015b) includes a more conservative table that includes code cycle dates that are considered to be equivalent to current codes and is included here as Table AE-2. This table provides the list of building codes presumed to be adequate if properly followed. School facility departments or the local building department may have information as to what code was used to design different facilities and to identify the "FEMA Building Type" of the facility.

Table AE-2   Building Codes Presumed to be Adequate (if Properly Followed) by FEMA P-154 (FEMA, 2015b)

	FEMA Building Type	Model Building Seismic Design Provisions		
		National Building Code/ Standard Building Code	Uniform Building Code	International Building Code
W1	Light wood frame single- or multiple-family dwellings of one or more stories in height	1993	1976	2000
W1A	Light wood frame multi-unit, multi-story residential buildings with plan areas on each floor of greater than 3.000 square feet	[1]	1997	2000
W2	Wood frame commercial and industrial buildings with a floor area larger than 5,000 square feet	1993	1976	2000
S1	Steel moment-resisting frame buildings	[1]	1994[2]	2000
S2	Braced steel frame buildings	[1]	1997	2000
S3	Light metal buildings	[1]	[1]	2000
S4	Steel frame buildings with concrete shear walls	1993	1994	2000
S5	Steel frame buildings with unreinforced masonry infill walls	[1]	[1]	2000
C1	Concrete moment-resisting frame buildings	1993	1994	2000
C2	Concrete shear wall buildings	1993	1994	2000
C3	Concrete frame buildings with unreinforced masonry infill walls	[1]	[1]	2000
PC1	Tilt-up buildings	[1]	1997	2000
PC2	Precast concrete frame buildings	[1]	[1]	2000
RM1	Reinforced masonry buildings with flexible floor and roof diaphragms	[1]	1997	2000
RM2	Reinforced masonry buildings with rigid floor and roof diaphragms	1993	1994	2000
URM	Unreinforced masonry bearing wall buildings	[1]	[1]	[1]
MH	Manufactured housing	[3]	[3]	[3]

[1]   No benchmark year.

[2]   Steel moment-resisting frame shall comply with the 1994 UBC Emergency Provisions, published September/October 1994.

[3]   The model building codes in this table do not apply to manufactured housing. In California, relevant requirements appeared in the Mobile Home Parks Act, the California Health and Safety Code, and the California Code of Regulations. They evolved between 1985 and 1994; the year 1995 is recommended here as the benchmark year for California. In other states, the U.S. Department of Housing and Urban Development's Installation Standards required tie-downs after October 2008. The year 2009 is recommended here as the benchmark year for states other than California.

Appendix F

Flood Maps Appendix

The *Flood Supplement* provided basic information about flood hazard maps, specifically, about FEMA's Flood Insurance Rate Map (FIRM). This appendix provides additional details about FIRMs, and examples of additional flood hazard maps that may be useful in evaluating existing school facilities and potential sites for new schools.

AF.1   Understanding and Using Flood Hazard Maps

Flood hazard maps can vary widely in terms of purpose. Not all maps are the same, but they can all provide useful information to stakeholders, including administrators, emergency managers, planners, designers, and teachers. Stakeholders should seek out and use as many flood hazard maps and as much flood hazard information as is available. This may mean using different maps for the same site.

Extracting and understanding all the information the maps and underlying studies contain may require assistance from others who specialize in flood, such as:

- Community planning or building departments;

- State emergency management agencies;

- State National Flood Insurance Program (NFIP) Coordinator's offices;

- State Hazard Mitigation Officers;

- Regional and state water resources agencies;

- Federal agencies (e.g., U.S. Geological Survey, U.S. Army Corps of Engineers, National Atmospheric and Oceanic Administration, Natural Resources, Conservation Service, and Federal Emergency Management Agency); and

- University departments, private companies and professional associations with expertise in hydrology, hydraulics, water resources, and flood hazard mapping.

Even with the assistance of groups like these, decisions may not be simple due to the complexity of flooding and flood maps. For this reason, it is helpful to understand basic information about flood hazard maps.

FEMA P-1000   **AF: Flood Maps Appendix**   **AF-1**

Flood Hazard Maps: Understanding Their Purpose, Assumptions, and Limitations

- Some flood hazard maps are intended to be used for evacuation planning (e.g., hurricane storm surge, tsunami, and dam break) or as general planning tools, while others are intended to be used at the parcel level for regulating design and construction of buildings.

- All flood hazard maps are based on some prescribed level of flood hazard, but the hazard level shown can vary from map to map, and the flood hazard level shown may or may not be adequate for life safety and property protection decisions.

- It is for this reason that building codes and regulations often modify the flood hazard or flood level shown on the map for building design or evaluation purposes. Flood maps should not be used alone; they should be used in concert with other applicable building regulations.

- Some maps are based on recent topography and flow calculations, while others may be based on data and calculation methods that are decades old. Maps based on topography, data, and/or methods that are outdated will not be as accurate.

- Some maps are the result of approximate studies, while others are based on detailed studies. The resolution and accuracy of the detailed studies will be better.

- Some maps consider flooding in small watersheds and streams, while others have a minimum watershed/stream size threshold required for flood calculations and mapping. Just because an area is shown as being free of flooding does not necessarily mean that is always the case.

- Some maps assume levees or flood protection structures or dams are strong enough to withstand a certain level of flooding and provide protection to nearby buildings, while some maps assume they are not. The true flood risk (including failure of a levee or similar structure) may not be indicated on any given flood hazard map.

- Some maps depict present-day flood hazards, while others project future flood hazards (due to increased development, changes in precipitation patterns, lake level rise, sea level rise, or erosion). Having estimates of future flood hazards improves our ability to forecast potential flood effects and flood damages over the life of a school.

AF.2 FIRMs and Related Products

A FIRM is what is called a "regulatory product." It is used to regulate development in flood hazard areas, to determine which buildings are required to carry flood insurance, and to determine the appropriate flood insurance rate for a building.

In almost all cases where a school site is being evaluated, a FIRM will be available. FIRMs are one of the products of a Flood Insurance Study (FIS). An FIS report will be published by FEMA with the FIRM, and the FIS will have detailed information and background that is not displayed on the map.

AF.2.1 Finding FIRMs and Related Products

FIRMs and related products can be obtained from many sources, but the two best sources will usually be: (1) FEMA's Map Service Center (MSC); and (2) the local or state government regulating floodplains.

The MSC (https://msc.fema.gov/portal/) allows users to search for flood hazard information by street address or by state/county/community (https://msc.fema.gov/portal/advanceSearch), or by other identifying information. The MSC provides the following types of products:

- Effective flood maps and studies—those adopted by jurisdictions for floodplain management and used by FEMA for flood insurance rating.

- Historic flood maps and studies—maps and studies that are no longer effective.

- Preliminary flood maps and studies—produced by a new flood study but not yet used by FEMA for flood insurance rating.

- Flood risk products—reports and maps that can be used for informational and planning purposes; availability is limited to the newest flood studies, and the exact products vary by location.

- GIS and database files from the National Flood Hazard Layer (NFHL), a digital nationwide compilation of effective FIRMs.

The NFHL is also accessible directly and can be viewed in web browsers without GIS software. To access this resource, visit http://fema.maps.arcgis .com/home/index.html, click on image above "flooding", and click on "FEMA's National Flood Hazard Layer (Official)."

Figures AF-1 through AF-3 provide screenshots of an example school used to demonstrate the use of MSC and NFHL sites. This school, Tanglewood Elementary School (9352 Rustling Oaks Avenue, Baton Rouge, Louisiana 70818), was flooded during the August 2016 Louisiana flood.

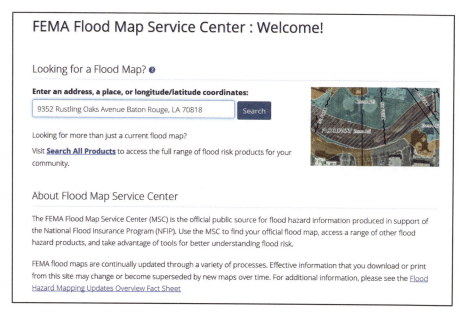

Figure AF-1   Screenshot from FEMA MSC website with the street address of Tanglewood Elementary School entered.

Figure AF-2   Screenshot showing MSC products available for Tanglewood Elementary School. Click on "view map" to see FIRM. Click on "interactive map" to see NFHL.

Figure AF-3   Portion of screenshot of NFHL image of Tanglewood Elementary School.

AF.2.2 Reading FIRMs

The exact map format may vary, but all FIRMs will show flood hazard zones, and most FIRMs will show Base Flood Elevations (BFEs)—the expected elevation of the flood surface during the base flood (also known as the 1% annual chance flood, or the 100-year flood). Some older FIRMs in less populated areas may show "approximate" flood zones without BFEs.

Table AF-1 lists the flood hazard zones shown on the FIRM (not all FIRMs will have all the zones listed in the table).

Figure AF-4 shows a generalized depiction of a FIRM to illustrate key information and terminology. Modern FIRMs are digital products drawn on aerial photographs (older FIRMs were drawn on paper maps).

FIRMs will show the following information:

- **Special Flood Hazard Area (SFHA).** This indicates the area expected to flood during the base flood, also known as the 1% annual chance (100-year) flood. FEMA considers the entire SFHA to be at high risk from flooding.

- **Base Flood Elevation (BFE).** This refers to the flood level associated with the Base Flood. The BFE will be listed on most FIRMs. The BFE will indicate the flood water surface elevation in riverine and lake flooding, and the elevation of the top of the wave in coastal flooding. Elevations will be relative to a common survey reference (vertical

> A tutorial on reading and understanding the FIRM is provided here https://www.fema.gov/media-library/assets/documents/7984?id=2324

datum). Flood elevation should not be confused with flood depth above ground.

Table AF-1 List of Flood Hazard Zones that May Be Shown on a FIRM (adapted from FEMA, 2005b)

Zone	Flood Hazard
Zone A	The 100-year or base floodplain. There are six types of A Zones: **A** — The base floodplain mapped by approximate methods, i.e., BFEs are not determined. This is often called an unnumbered A Zone or an approximate A Zone. **A1-30** — These are known as numbered A Zones (e.g., A7 or A14). This is the base floodplain where FIRM shows a BFE (old format). **AE** — The base floodplain where base flood elevations are provided. AE Zones are now used on new format FIRMS instead of A1-A30 Zones. **AO** — The base floodplain with sheet flow, ponding, or shallow flooding. Base flood depths (feet above ground) are provided. **AH** — Shallow flooding base floodplain. BFEs are provided. **A99** — Area to be protected from base flood by levees or Federal Flood Protection Systems under construction. BFEs are not determined. **AR** — The base floodplain that results from the decertification of a previously accredited flood protection system that is in the process of being restored to provide a 100-year or greater level of flood protection.
Zone V and Zone VE	**V** — The coastal area subject to a velocity hazard (wave action) where BFEs are not determined on the FIRM. **VE** — The coastal area subject to a velocity hazard (wave action) where BFEs are provided on the FIRM.
Zone B and Zone X (shaded)	Area of moderate flood hazard, usually the area between the limits of the 100-year and 500-year floods. B Zones are also used to designate base floodplains of lesser hazards, such as areas protected by levees from the 100-year flood, or shallow flooding areas with average depths of less than one foot or drainage areas less than 1 square mile.
Zone C and Zone X (unshaded)	Area of minimal flood hazard, usually depicted on FIRMs as above the 500-year flood level. Zone C may have ponding and local drainage problems that don't warrant a detailed study or designation as a base floodplain. Zone X is the area determined to be outside the 500-year flood and protected by levee from 100-year flood.
Zone D	Area of undetermined but possible flood hazards.

- **Zones.** The SFHA will be subdivided into zones, and the zone designations will vary with type of flood hazard area (riverine or coastal), age of the FIRM, and the level of detail of the flood study that created the FIRM.

  o The SFHA has mostly "A Zones" and "V Zones" (exact letter designations may vary).

  o V Zones are mapped along the shoreline in coastal areas and indicate the presence of damaging waves during the base flood.

- A Zones are mapped for riverine, lake and coastal flooding, and indicate inundation (small waves may be present in coastal).
- The "floodway" is an SFHA overlay that may be mapped in rivers and large streams. The floodway is an area subject to the highest riverine flood hazards, and where special development restrictions exist.

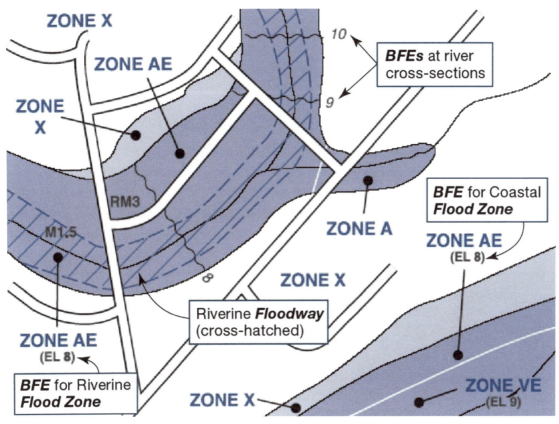

Figure AF-4 Generalized depiction of a FEMA Flood Insurance Rate Map (FIRM), showing flood hazard zones and Base Flood Elevations (BFEs). A riverine flood hazard area is shown at the upper left, and a coastal flood hazard area is shown at the lower right.

- Many FIRMs will show the area expected to flood in a more severe flood, usually the 0.2% annual chance (500-year) flood. This area will be shown as Zone X (or B or C). BFEs are not mapped for these zones, but flood elevation information may be available in the FIS.

- Newer coastal FIRMs will show a line representing the Limit of Moderate Wave Action (LiMWA), which is used by building codes and standards to identify the Coastal A Zone, within which those codes and standards require design and construction to V Zone standards.

FEMA's Risk Mapping, Assessment, and Planning (Risk MAP) program also develops "non-regulatory" products that are informational in nature (https://www.ready.gov/tsunamis). These products may include flood risk reports

> To see if Risk MAP non-regulatory products are available for your area, look on the Map Service Center web site, or contact your community floodplain manager.

and databases, and special maps that show flood depths and velocities, anticipated financial losses from flood, and areas of mitigation interest.

AF.3 Other Types of Flood Hazard Maps

As was stated previously, FIRMs are not the only type of flood hazard map that should be consulted when flood vulnerability of a school is evaluated. The following sections compare other types of flood maps with the FIRM.

AF.3.1 Coastal Flood Hazard Map Comparison

Figures AF-5 and AF-6 show different flood hazard mapping approaches for the same general area of Tampa, Florida.

- Figure AF-5 shows the NFHL version of the FIRM produced by FEMA.

- Figure AF-6 shows a storm surge inundation map produced by the Tampa Bay Regional Planning Council and the State of Florida during a hurricane evacuation study. Inundation maps like this serve as the basis for community evacuation plans. Some states and communities may establish additional siting restrictions or design requirements using storm surge inundation maps.

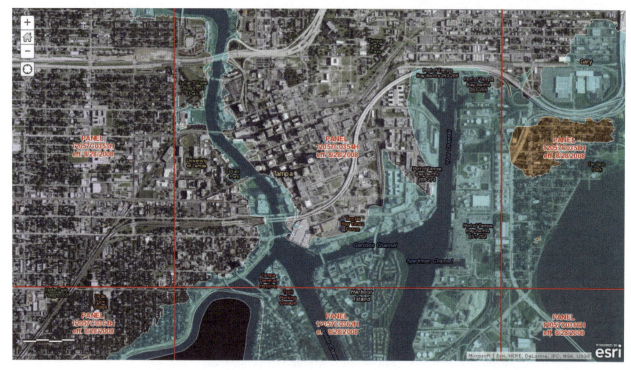

Figure AF-5   Portion of screenshot of the portion of the National Flood Hazard Layer for Tampa, Florida. Cyan shading shows the area expected to flood during the 1% annual chance flood.

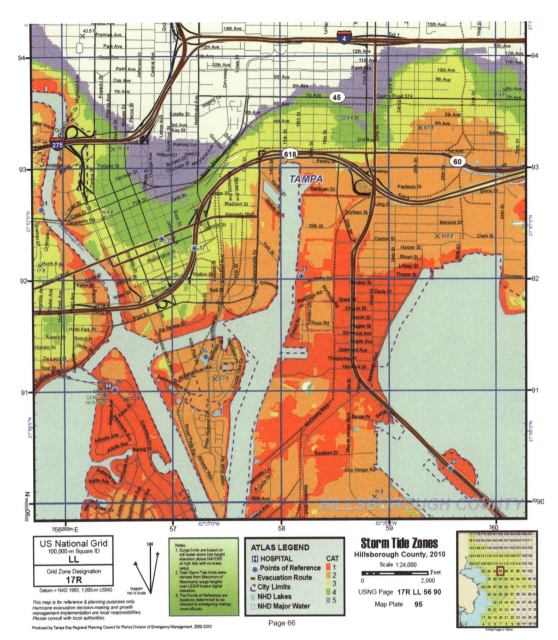

Figure AF-6 Storm surge atlas for Tampa, Florida (including the area shown in Figure AF-5). Shading color indicates areas expected to flood during different strength hurricanes, from red for Category 1 (minimal hurricane) to lavender for Category 5 (most intense hurricane). (Photo source: Tampa Bay Regional Planning Council)

It is apparent by comparing Figures AF-5 and AF-6, that a given location may be outside the 1% annual chance flood hazard area (which is subject to floodplain regulations and flood-resistant design requirements of building codes), but inside a storm surge inundation area. Both these maps were produced for different purposes, so their use may not be interchangeable. However, both can inform flood hazard evaluations and school siting and design decisions.

> Dam inundation areas have not been mapped everywhere, and some maps are not available to the public. For more information, contact the state dam safety program or visit: https://www.fema.gov/about-national-dam-safety-program.

AF.3.2 Riverine Flood Hazard Map Comparison

Another flood map comparison example is shown in Figures AF-7 and AF-8, this comparison for a different type of inundation hazard—dam failure. Figure AF-7 shows the FIRM for an area near Lawton, Oklahoma. Figure AF-8 shows the corresponding area that would be subject to inundation if the dam at Lake Lawtonka failed.

We see from Figures AF-7 and AF-8 that a given location may be outside the 1% annual chance flood hazard area (which is subject to floodplain regulations and flood-resistant design requirements of building codes), but inside a dam failure inundation area. As stated previously, although produced for different reasons, both of these maps should be used for flood hazard evaluations and school design decisions in dam inundation areas.

Figure AF-7   Portion of screenshot from the National Flood Hazard Layer for Lawton, Oklahoma. For comparison with Figure AF-8, the red circle is at a running track at Pritchard Field, outside the area expected to flood during the 1% annual chance flood – cyan shading).

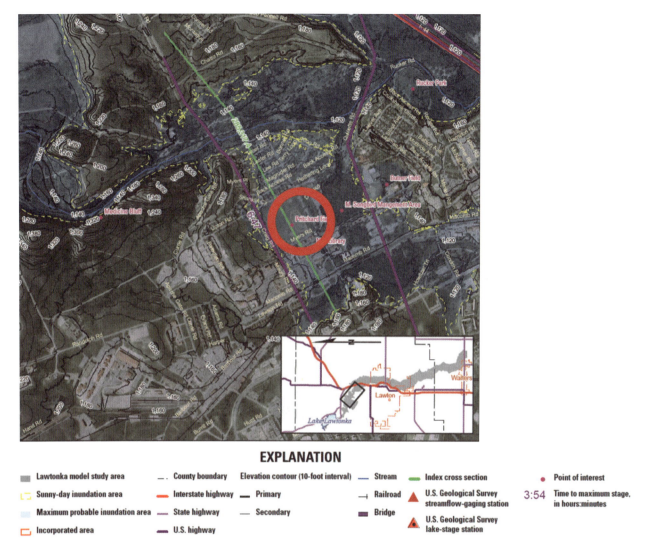

Figure AF-8   Dam failure inundation map for the area shown in Figure AF-7 (USGS, 2012). The area in the red circle (at a track at Pritchard Field) is subject to inundation during modeled dam failure scenarios (cyan shading and yellow dashed line).

AF.3.3 Future Conditions Affecting Flood Hazards

FIRMs and most other flood hazard maps are almost always based on existing conditions, meaning flood conditions at the time the flood map was created. Maps should be updated as conditions change, but unfortunately, most map updates lag behind the actual change. Evaluation of school sites should include future conditions flood hazard conditions, where possible.

Flood conditions can change over time for any number of reasons, including:

- Future development and land use can increase the amount of impervious ground surface or speed rainfall runoff, affecting flood timing and elevations.

- Surface water, lake, and groundwater levels can change over time, and these can affect future flooding.

- Precipitation patterns may change over time, increasing flood levels in certain areas.

- Sea level can increase over time, increasing coastal flood levels.

- Erosion can occur along coasts, rivers, streams, and lakes, and the area subject to flooding will change as the shoreline moves.

Figure AF-9 shows an example where a community (Charlotte-Mecklenburg, North Carolina) produced two sets of flood hazards on their FIRM, one based on existing conditions (FEMA approach), and one showing the expected flood hazard area based on future development build-out conditions.

Figure AF-9 Charlotte-Mecklenburg, North Carolina, flood hazard map showing 1% annual chance flood hazard area (blue shaded area) and community flood hazard area based on future development upstream (gray shaded area). (Photo source: City of Charlotte)

There are now many tools and maps that predict inundation as a result of sea level rise (e.g., http://gom.usgs.gov/slr/slr.aspx, https://coast.noaa.gov/slr/, and http://sealevel.climatecentral.org/maps/risk-zone), but most of these show sea level rise inundation scenarios under normal tide conditions, or rely on the user to estimate a combined storm surge and sea level rise; they do not show future flood hazard areas that include sea level rise effects layered on top of the FIRM.

In some cases, tools and maps combine sea level rise and storm inundation; these will be the most useful for evaluating potential school sites. Figure AF-10 shows a sample map that displays future sea level rise effects in conjunction with the current 1% annual chance (100-yr) flood hazard area.

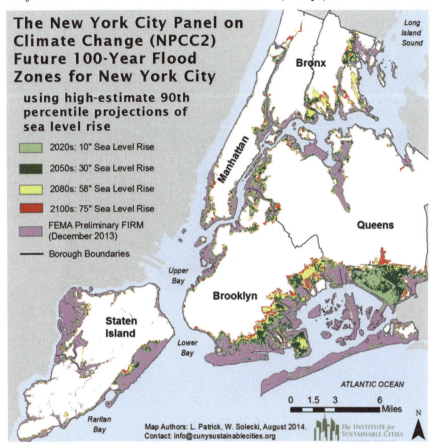

Figure AF-10   Areas that potentially could be impacted by the 100-yr flood and various sea level rise heights (Patrick at al., 2015).

FEMA has initiated several pilot studies examining sea level rise impacts on mapped flood hazards, and has incorporated "increased flooding scenarios" (e.g., BFE + 1 ft, BFE + 2 ft, and BFE + 3 ft) into its Risk MAP products for some studies. Figure AF-11 shows a brochure for this mapping product for the San Francisco Bay Area.

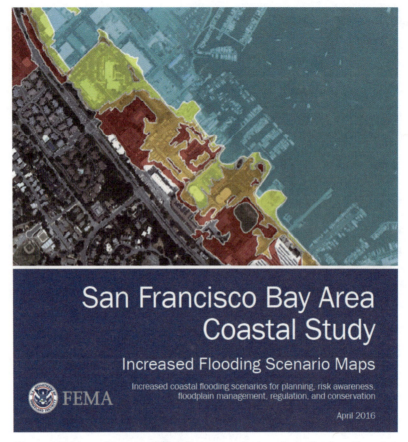

Figure AF-11   Increased flooding scenario brochure for San Francisco Bay Area (FEMA, 2016).

Appendix AR

Resources

Full citations of all references used to develop this *Guide* are listed in the References. The following is a list of recommended resources that might be useful for school leaders that are addressing natural hazard risk. In some cases, a document is both in the References and listed here as a recommended resource. For recommended resources for a particular hazard, see the recommended resources section in each hazard-specific supplement.

AR.1 Communications

- Excellent resources for developing messages, school crisis communications plans, and offering helpful tips on crisis communications and planning are available for purchase and for free online. Resource include:

  o *The Complete Crisis Communication Management Manual for Schools* (NSPRA, 2016) published by the National School Public Relations Association, available for purchase. http://www.nspra.org /crisis

  o *School Crisis Guide: Help and Healing in a Time of Crisis* (National Education Association, 2015). This step-by-step resource is created by educators for educators. http://healthyfutures.nea.org/wpcproduct /school-crisis-guide/

- *Public Awareness and Public Education for Disaster Risk Reduction: A Guide* (IFRC, 2011). This document provides guidance on planning and developing public awareness and public education efforts for disaster risk reduction. http://www.ifrc.org/Global/Publications/disasters /reducing_risks/302200-Public-awareness-DDR-guide-EN.pdf

- Online resources to learn more about the use of social media in emergency management include:

  o Virtual EMA, a professional association of emergency responders, academia, technologists, community advocates, creative thinkers with educational resource links on use of social media for emergency management. http://www.virtualema.org (formerly http://www .sm4em.org)

FEMA P-1000   AR: Resources   AR-1

- FEMA resource page for social media; https://www.fema.gov/social-media. Additionally, FEMA offers a series of courses instructing schools, first responders, and other organizations on using social media in disaster response and recovery. Courses are offered free of charge through the National Disaster Preparedness Training Center (https://ndptc.hawaii.edu/training/catalog).

AR.2 Community Engagement

Teen Community Emergency Response Team (CERT) is an excellent program to involve youth. To learn more about building CERT teams for schools, visit:

- https://rems.ed.gov/TeenCertEnhancingSchoolEmergMgrment.aspx

- http://www.fema.gov/teen-community-emergency-response-team

The following is a program for knowledge-sharing around disasters:

- Voluntary Agencies Active in Disaster (VOAD). https://www.ready.gov/voluntary-organizations-active-disaster

AR.3 Curriculum

- Many examples of educational curriculum used throughout the world are readily available online. Four excellent examples are:

  - *What's the Plan, Stan?* is an initiative from New Zealand, which aims to support teachers to develop their students' knowledge, skills and attitudes to respond to and prepare for an emergency. http://www.whatstheplanstan.govt.nz

  - The Pillowcase Project. The American Red Cross, sponsored by Disney, is a preparedness education program for children in grades 3–5, which teaches students about personal and family preparedness, local hazards, and basic coping skills. The hour-long presentation is for schools, after-school programs, Girl and Boy Scout troop meetings, and more. Red Cross volunteers lead students through a "learn, practice, share" framework to engage them in disaster preparedness. Upon completion, students receive a sturdy pillowcase in which to build their personal emergency supplies kit. Presentations are customized to focus on a hazard that is important and relevant to the local community. http://www.redcross.org/local/colorado/programs-services/ preparedness/pillowcase-project

  - Masters of Disaster. The American Red Cross Masters of Disaster® curriculum features ready-to-go lesson plans that help educate children about important disaster safety and preparedness information. The Masters of Disaster curriculum materials are specifically tailored for lower elementary (K–2), upper elementary (3–5) and middle school (6–8) classes. http://www.redcross.org /local/utah/programs-services/masters-of-disaster-program

- o Be A Hero. This FEMA program includes online activities for children to learn about and prepare for disasters. Curriculum is also provided by grade level for educators to utilize in the classroom and are designed to be engaging, interactive, and multidisciplinary. www.ready.gov/kids

- Educational resources:

  - o riskRED: risk reduction education for disasters. http://www.riskred .org/schools.html

  - o Edu4drr: Effective Education for Disaster Risk Reduction-Learning Matters website. http://www.edu4drr.org

  - o Drrlibrary (resources). http://www.drrlibrary.org/about.php

  - o PreventionWeb (International disaster resilience networks, coalitions and resources). http://www.preventionweb.net/english/

  - o Coalition for Global School Safety and Disaster Prevention Education. http://cogssdpe.ning.com

  - o *Disaster Risk Reduction in School Curricula: Case Studies from Thirty Countries* (Selby and Kagawa, 2012). This publication provides best practices and innovative solutions throughout the world with the intention of providing support to all countries in the process of integrating disaster risk reduction into curriculum. https://www.unicef.org/education/files/DRRinCurricula-Mapping 30countriesFINAL.pdf

  - o *Children and Disaster Risk Reduction: Taking Stock and Moving Forward* (Back et al., 2009). This report reviews child-focused and child-led disaster risk reduction approaches and techniques. It documents a number of case studies across a range of interventions, dividing these into three main areas: Knowledge, Voice and Action. It makes some observations regarding current practice and recommendations that imply a shift in emphasis going forward. http://toolkit.ineesite.org/toolkit/INEEcms/uploads/1057/Children _and_Disaster_Risk_Reduction.pdf

- *Towards a Learning Culture of Safety and Resilience: Technical Guidance for Integrating Disaster Risk Reduction in the School Curriculum* (Selby and Kagawa, 2014). This document provides a thorough rationale for including disaster risk reduction in school curricula within an Education for Sustainable Development framework. It also guides those with responsibility for curricula on appropriate teaching and learning methods for disaster preparedness. http://unesdoc.unesco.org/images/ 0022/002293/229336E.pdf

  - PrepareAthon! America's PrepareAthon! is a grassroots campaign for action to increase community preparedness and resilience. Join others around the country to practice your preparedness! https:// community.fema.gov/

  - National Weather Service Education. An educational resource for children to learn about science and safety with Owlie Skywarn. http://www.weather.gov/owlie/

  - National Oceanic and Atmospheric Administration (NOAA) Education Resources, Natural Disasters. The resources on this page provide background information on the types of natural hazards that NOAA studies, techniques and tools used for prediction, historical patterns of natural disasters, and programs for helping communities prepare for future events. http://www.noaa.gov/resource-collections /weather-atmosphere-education-resources

  - STOP DISASTERS! A disaster simulation game from the UNISDR that teaches principles of disaster preparedness for earthquakes, floods, cyclones, and wildfires. www.stop disastersgame.org

- "Big Bird, Disaster Masters, and High School Students Taking Charge: The Social Capacities of Children in Disaster Education" (Wachtendorf et al., 2008). This paper reviews three initiatives that focus on children and disasters, including a Sesame Workshop-produced video aimed at pre-school children, an American Red Cross initiative that focuses on children in kindergarten through middle school, and a video directed at high school students as part of a student-generated initiative at a Seattle school.

AR.4   Disaster and Emergency Planning

AR.4.1 General Guidance

- *Guide for Developing High-Quality School Emergency Operations Plans* (U.S. Department of Education, 2013) provides recommendations for developing plans to respond to an emergency and also outlines how K-12 schools can plan to prevent, protect against, mitigate impacts of, and recover from these emergencies. https://www.fema.gov/media-library/assets/documents/33599

- The federally funded Readiness and Emergency Management Schools Technical Assistance (REMS TA) Center, administered by the U.S. Department of Education, Office of Safe and Healthy Students, provides extensive online resources including development guidelines, interactive worksheets, plan templates, a toolbox, trainings, and other resources to assist in the process. https://rems.ed.gov/

- Go-Kits and Emergency Supplies. Sources of information on Go-Kit basics, including recommended contents and sample checklists, include the following links:

  o REMS "Helpful Hints" Newsletter (2006). http://rems.ed.gov/docs/HH_Vol1Issue1.pdf

  o Washington (State) School Safety Center. http://www.k12.wa.us/Safetycenter/Emergency/pubdocs/EmergencySuppliesGoKitSuggestions.pdf

- Multihazard Emergency Planning for Schools Toolkit. This is an online resource to support emergency planning efforts. https://training.fema.gov/programs/emischool/el361toolkit/start.htm

- FEMA Course IS-362: *Multi-Hazard Emergency Planning for Schools*. This online course covers basic information about developing, implementing, and maintaining school emergency operations plan (EOP), with the goals of providing an understanding of the importance of having an EOP for schools. This course is intended for teachers, counselors, parent volunteers, coaches, bus drivers, and students; school administrators, principals, and first responders might also find the information useful. https://training.fema.gov/is/courseoverview.aspx?code=IS-362.a

- FEMA Course E361: *Multi-Hazard Emergency Planning for Schools*. This 4-day in-person course provides school district teams with the knowledge, skills and tools needed to review, enhance and sustain an all-hazard school emergency plan (EOP). https://training.fema.gov/emicourses/crsdetail.aspx?cid=E361&ctype=R

- *Disaster and Emergency Preparedness: Guidance for Schools* (International Finance Corporation, 2010). This handbook was prepared as a resource for school administrators and teachers to serve as a basis for policy development. It also provides an important resource for classroom activities and awareness-raising among children and communities. This guide includes an addenda of checklists useful for school emergency planning. http://www.ifc.org/wps/wcm/connect/8b796b004970c0199a7ada336b93d75f/DisERHandbook.pdf?MOD=AJPERES. *The Disaster and Emergency Preparedness: Activity Guide for K to 6th Grade Teachers*. This acts as a supplemental guide to the handbook. http://toolkit.ineesite.org/resources/ineecms/uploads/1057/Disaster_Emerg_Preparedness_K-6.pdf

- *Practical Information on Crisis Planning: A Guide for Schools and Communities* (U.S. Department of Education, 2007). This guide provides a thorough overview of crisis planning for schools, including natural disaster planning. The focus of this document is on the cycle of crisis management and is intended to be a general guide. It includes action steps for each stage of planning and can be modified for any school or type of crisis. http://www2.ed.gov/admins/lead/safety/emergencyplan/ crisisplanning.pdf

- "Creating emergency management plans" (U.S. Department of Education, 2006). This document provides brief guidance for emergency planning for schools with specific and useful suggestions. http://rems.ed.gov/docs/creatingplans.pdf

- *Disaster Prevention for Schools: Guidance for Education Sector Decision-Makers* (Petal, 2008). This is a guidance document for school administrators, teachers, education authorities, school safety committees and similar. It introduces disaster impacts on and prevention for schools, and covers creating and maintaining safe learning environments; teaching and learning disaster prevention and preparedness; educational materials and teacher training; and developing a culture of safety. http://www.preventionweb.net/educational/view/7344

- *Head Start: Disaster Preparedness Workbook* (UCLA Center for Public Health and Disasters, 2004). This workbook is specific to head start programs and provides an in-depth look at disaster planning with specific suggestions, activities, and helpful checklists, worksheets, and forms to prepare your head start program. It includes great information on training of staff, communication systems, creating response teams, and recovery. http://toolkit.ineesite.org/toolkit/INEEcms/uploads/1056/Headstart_Disaster_Preparedness.pdf

- i love u guys foundation. This organization provides resources for developing standard response protocols for use in any emergency situation. The foundation provides training, forms, checklists, and additional resources. www.iloveuguys.org

- Center for Safe Schools Emergency Response Crisis Management (ERCM) Initiative. This center guides schools and districts through a 4-step process of prevention/mitigation, preparedness, response, and recovery; and provides them with needed research, training and consultation. http://www.safeschools.info/emergency-management

- Safe School Facilities Checklist. The National Clearinghouse for Educational Facilities (NCEF) provides a free checklist that combines the nation's best school facility assessment measures into one online source for assessing the safety and security of school buildings and grounds. http://www.ncef.org/content/safe-schools-0

- *School Safety Plans: A Snapshot of Legislative Action* (Council of State Governments Justice Center, 2014). This CSG Justice Center brief recaps the legislation of 46 states with respect to their requirements for comprehensive school safety or emergency plans. https://csgjustice center.org/youth/publications/ school-safety-plans-a-snapshot-of -legislative-action/

AR.4.2 Federal Laws Applicable to Emergency Operations Plans

- Information on applicable federal laws is discussed in *Guide for Developing High Quality School Emergency Operations Plans* (US Department of Education, et al., 2013). More resources on the following are available at:

  - Americans with Disabilities Act. http://www.ada.gov and http://achieve.lausd.net/afn

  - Title VI, Civil Rights Act of 1964. https://www.justice.gov/crt/fcs /TitleVI-Overview

  - Family Educational Rights and Privacy Act (FERPA). http://www2 .ed.gov/policy/gen/guid/fpco/ferpa/index.html

  - *Emergency Management: Status of School Districts' Planning and Preparedness* (U.S. General Accountability Office, 2007). http:// www.gao.gov/products/GAO-07-821T

  - The Stafford Act. http://www.fema.gov/robert-t-stafford-disaster -relief-and-emergency-assistance-act-public-law-93-288-amended

AR.4.3 Incident Command System

- The online ICS Resource Center from the FEMA Emergency Management Institute. This website provides training in ICS, useful forms, and job aids. https://training.fema.gov/emiweb/is/icsresource/

- Independent Study courses are available on FEMA's website at http://training.fema.gov/emi.aspx. Courses include instruction in the National Institute Management System (NIMS) and the Incident Command System (ICS).

- National Incident Management System (NIMS) is a systematic, proactive approach to guide departments and agencies at all levels of government, nongovernmental organizations, and the private sector to work together seamlessly and manage incidents involving all threats and hazards. http://www.fema.gov/national-incident-management-system

AR.4.4 School Safety Plan Examples

- *Crisis Management and Prevention Information for Georgia Public Schools* (Georgia Department of Education, 2012). This manual of information was developed to support and encourage the development and implementation of a systematic crisis management and prevention plan in schools and school districts. It is designed to be used as a general resource and a training tool. It is not intended to list or discuss all possible emergency situations or conditions in a school setting. https://www.gadoe.org/Curriculum-Instruction-and-Assessment /Curriculum-and-Instruction/Documents/Crisis%20Management %20and%20Prevention%20in%20Georgia%20Public%20Schools _December%202012.pdf

- This workbook from Fairfax County Public Schools includes excellent checklists and includes all of the necessary requirements for a thorough school plan. http://www.esc1.net/cms/lib/TX21000366/Centricity /Domain/89/Crisis_Management_Workbook_Fairfax_County_Public _Schools.pdf

- *Emergency Preparedness Planning Guide for Utah Public Schools* (Utah State Office of Education, 2013). This plan can be found at the be ready website which also contains many additional useful resources for school emergency planning. www.beready.utah.gov

- This resource guide from the Kentucky Center for School Safety can be used as a template that can be adjusted as needed. http://kycss.org/emp /Home/EmerRevCol.pdf. Additional resources: www.kysafeschools.org

- *Washington State K-12 Facilities Hazard Mitigation Plan* (School Facilities and Organization, 2014). This document can be used as a template for other states to develop similar hazard plans. Specific disasters discussed include earthquakes, tsunamis, volcanic eruption, flooding, and landslides. http://www.k12.wa.us/SchFacilities/PDM /pubdocs/PDM_Plan.pdf

- *Emergency Operations Planning: A Model for Schools and Businesses* (Tennessee Office of Homeland Security, 2014). This guide uses a simplified model of emergency protocols for specific incidences. This template can be used as a quick reference guide for any school to follow. https://www.tn.gov/assets/entities/education/attachments/save_act _emergency_ops_planning_model.pdf

- *Sample School Emergency Operations Plan* (FEMA, 2013b). This document is designed to be used alongside the *Guide for Developing High Quality School Emergency Operations Plan* (U.S. Department of Education, 2013). It presents excepts from a sample school emergency plan that can be modified as needed. https://www.educateiowa.gov /documents/school-safety/2016/02/sample-school-emergency-operations -plan-fema-nov-2013

AR.5 Emergency Exercises, Drills, and Materials

- *Homeland Security Exercise and Evaluation Program (HSEEP)* (Homeland Security, 2013). This document provides detailed approaches, strategies and tips on how to successfully design, execute, and evaluate an exercise. https://www.fema.gov/media-library/assets /documents/32326

- "Accountability and assessment of emergency drill performance at schools" (Ramirez et al., 2009). This mixed-methods study measures the quantity and the quality of drills in an urban school district in Los Angeles. Suggestions include developing realistic simulated exercises, debriefing, and better school accountability for drills. https://www.ncbi .nlm.nih.gov/pubmed/19305209

AR.6 International Resources

- *Comprehensive School Safety: A Global Framework.* (GADRRRES, 2017). This framework provides a comprehensive approach to reducing risks from all hazards to the education sector. The purpose of this Comprehensive School Safety Framework is to bring these efforts into a clear and unified focus in order for education sector partners to work more effectively, and to link with similar efforts in all other sectors at the global, regional, national and local levels. http://gadrrres.net/resources/comprehensive-school-safety-framework

- *School Recovery: Lessons from Asia* (Shaw et al., 2012). This work was conducted at the Graduate School of Global Environmental Studies at Kyoto University under International Environment and Disaster Management. This publication provides a compilation of 25 case studies of schools from 12 countries affected by 6 different hazards in an effort to understand the recovery processes of schools from natural disasters. It links the recovery lessons to the Hyogo Framework for Action (HFA) in the education sector (E-HFA) through sixteen tasks that constitute a framework for the integrated approach of disaster risk reduction in schools. It goes beyond the school building or education, and looks at the comprehensive way of disaster risk reduction in the education sector.

- *Guidance Notes on Safer School Construction* (GFDRR, 2009). This document presents a framework of guiding principles and general steps to develop a context-specific plan to address this critical gap to reaching Education for All (EFA) and Millennium Development Goals (MDGs) through the disaster resilient construction and retrofitting of school buildings. https://www.eeri.org/wp-content/uploads/Guidance_Notes_Safer_School_Constructionfinal.pdf

AR.7 Mental Health

- Training resources on Psychological First Aid for Schools (PFA-S) and other mental health resources are offered by The National Child Traumatic Stress Network. http://www.nctsnet.org

- The following is a presentation on best practices in mental health emergency recovery in schools: http://www.powershow.com/view/3b1fd0-NjAwN/Best_Practices_in_Mental_Health_Emergency_Recovery_in_Schools_powerpoint_ppt_presentation

- Additional guidance on mental health recovery can be found at:

  o http://neahealthyfutures.org/wpcproduct/school-crisis-guide/

  o http://www.nspra.org/crisis

  o http://www.schoolcrisisresponse.com

  o http://www.atsm.org

  o http://safesupportivelearning.ed.gov

  o http://www.dodea.edu/crisis/cMresources.cfm

- "Do It Now: Short Term Responses to Traumatic Events" (Demaria and Schonfeld, 2014). This source provides an overview of best practices for assisting children with coping following a disaster.

- *School-Wide Crisis Management Plan Guide: A Professional School Counselor's Guide to School-Wide Crisis Management* (Missouri Professional School Counselors and Counselor Educators, 2015). http://www.missouricareereducation.org/doc/schcrisis/SchoolwideCrisis.pdf

- "Helping Elementary-Age Children Cope with Disasters" (Shen and Sink, 2002). This article addresses the effects of disasters on elementary-age children and their needs for mental health. It suggests possible school-based interventions and provides a case study of a traumatized first-grader, demonstrating how child- centered play therapy can be used in school settings.

- "School Interventions After the Joplin Tornado" (Kanter and Abramson, 2013). This article provides helpful information on the use of post-disaster mental health services for school children.

- Tips for Talking with and Helping Children and Youth Cope After a Disaster or Traumatic Event: A Guide for Parents, Caregivers, and Teachers. (SAMHSA, 2012). This document aims to help parents and teachers recognize common reactions children of different age groups (preschool and early childhood to adolescence) experience after a disaster or traumatic event. It offers tips for how to respond in a helpful way and when to seek support. http://store.samhsa.gov/product/Tips-for-Talking-With-and-Helping-Children-and-Youth-Cope-After-a-Disaster-or-Traumatic-Event-A-Guide-for-Parents-Caregivers-and-Teachers/SMA12-4732

- "Pedagogy of Love and Care: Shaken Schools Respond" (O'Connor, 2013). The paper continues to look at a pedagogy of love and care as a vital way to support, encourage, and aid children through their trauma and grief which continues long after the earthquake.

- "'I Had to Teach Hard': Traumatic Conditions and Teachers in Post-Katrina Classrooms" (Alvarez, 2010).

- "Consequences for Classroom Environments and School Personnel: Evaluating Katrina's Effect on Schools and System Response" (Buchanan and Baumgartner, 2010).

AR.8 Vulnerability Assessments

- *A Guide to School Vulnerability Assessments: Key Principles for Safe Schools* (U.S. Department of Education, 2008). This is a companion document to *Practical Information on Crisis Planning a Guide for Schools and Communities* (U.S. Department of Education, 2007) and is intended to serve as a guide for schools and districts to prepare for a variety of crises. The document emphasizes ongoing vulnerability assessments and is intended to assist schools with the implementation of an effective vulnerability assessment process. http://www.prevention web.net/files/15318_vareport20081.pdf

References

Alexander, D., and Lewis, L., 2014, *Condition of America's Public School Facilities: 2012-13*, NCES 2014-022, National Center for Education Statistics, U.S. Department of Education, Washington, D.C.

Alvarez, D., 2010, "'I had to teach hard': Traumatic conditions and teachers in post-Katrina classrooms," *The High School Journal*, Vol. 94, No. 1, pp. 28-39.

ASCE, 2009, *So, You Live Behind a Levee!*, American Society of Civil Engineers, Reston, Virginia.

ASCE, 2010, *Minimum Design Loads for Buildings and Other Structures*, ASCE/SEI 7-10, American Society of Civil Engineers, Reston, Virginia.

ASCE, 2014a, *Flood Resistant Design and Construction*, ASCE/SEI 24-14, American Society of Civil Engineers, Reston, Virginia.

ASCE, 2014b, *Seismic Evaluation and Retrofit of Existing Buildings*, ASCE/SEI 41-13, American Society of Civil Engineers, Reston, Virginia.

ASCE, 2017a, *2017 Infrastructure Report Card*, American Society of Civil Engineers, Reston, Virginia.

ASCE, 2017b, *Minimum Design Loads and Associated Criteria for Buildings and Other Structures*, ASCE/SEI 7-16, American Society of Civil Engineers, Reston, Virginia.

ATC, 2004, *Field Manual: Safety Evaluation of Buildings after Windstorms and Floods*, Applied Technology Council, ATC-45 Report, Redwood City, California.

ATC, 2005, *Field Manual: Postearthquake Safety Evaluation of Buildings*, Second Edition, Applied Technology Council, ATC-20-1 Report, Redwood City, California.

Back, E., Cameron, C., and Tanner, T., 2009, *Children and Disaster Risk Reduction: Taking Stock and Moving Forward*, United Nations Children's Fund (UNICEF), New York, New York.

Bailey, B., 2013, "Local banks step in after tornado prevented Moore district from paying schoolteachers on time," *The Oklahoman*, http://newsok.com/article/3844466, last accessed on February 20, 2017.

Balassanian, S.Y., Arkakelian, A.R., Nazaretian, S.N., Avanessian, A.S., Martirossian, A.H., Igoumnov, V.A., Melkoumian, M.G., Manoukian, A.V., and Tovmassian, A.K., 1995, "Retrospective analysis of the Spitak earthquake," *Annali di Geofisica*, Vol. 38, No. 3-4, Bologna, Italy.

Barnes, J., 2013, *School Disaster Needs for Students with Disabilities: Voices from the Field*, University of California at Los Angeles, Los Angeles, California.

Buchanan, T.K., and Baumgartner, J.J., 2010, "Consequences for classroom environments and school personnel: evaluating Katrina's effect on schools and system response," *Helping Families and Communities Recover from Disaster: Lessons Learned from Hurricane Katrina and its Aftermath*, American Psychological Association, Washington, D.C.

California Emergency Management Agency, 2011, *Guide and Checklist for Nonstructural Earthquake Hazards in California Schools*, Division of State Architect, Mather, California.

Connections Academy, 2016, *Virtual Public Schools Expands Enrollment to Accommodate Families Impacted by Flood*, www.connections academy.com/news/laca-expands-enrollment, last accessed on February 20, 2017.

Council of State Governments Justice Center, 2014, *School Safety Plans: A Snapshot of Legislative Action*, Council of State Governments Justice Center, New York, New York, https://csgjusticecenter.org/wp -content/uploads/2014/03/NCSL-School-Safety-Plans-Brief.pdf, last accessed on June 11, 2017.

CSSC, 2007, *The Field Act and Public School Construction: A 2007 Perspective*, State of California Seismic Safety Commission, Sacramento, California.

CSSC, 2009, *The Field Act and its Relative Effectiveness in Reducing Earthquake Damage in California Public Schools*, State of California Seismic Safety Commission, Sacramento, California.

Demaria, T., and Schonfeld, D.J., 2014, "Do it now: Short term responses to traumatic events," *Phi Delta Kappan*, Vol. 95, No. 4, pp. 13-17.

Doughton, S., and Gilbert, D., 2016, "'We should be screaming' with outrage: State does little to protect schoolkids from earthquake, tsunami," *The Seattle Times*, Seattle, Washington,

Educational Facilities Research Center, 2005, *Case Studies of Seismic Nonstructural Retrofitting in School Facilities*, National Institute for Educational Policy Research, Japan.

FEMA, 1999, *Building Performance Assessment Team (BPAT) Report – Hurricane Georges in Puerto Rico (1999)*, FEMA 339, Federal Emergency Management Agency, Washington, D.C.

FEMA, 2000a, *Design and Construction Guidance for Community Shelters*, First Edition, FEMA 361, Federal Emergency Management Agency, Washington, D.C.

FEMA, 2000b, *Tremor Troop: Earthquakes – A Teacher's Package for K-6*, Revised Edition, FEMA 159, Federal Emergency Management Agency, Washington, D.C.

FEMA, 2002, *Mitigation Case Studies: Protecting School Children from Tornadoes, State of Kansas School Shelter Initiative*, Federal Emergency Management Agency, Washington, D.C.

FEMA, 2003, *Risk Management Series: Incremental Seismic Rehabilitation of School Buildings (K-12)*, FEMA 395, Federal Emergency Management Agency, Washington, D.C.

FEMA, 2005a, *Mitigation Assessment Team Report: Hurricane Charley in Florida: Observations, Recommendations, and Technical Guidance*, FEMA 488, Federal Emergency Management Agency, Washington, D.C.

FEMA, 2005b, *NFIP Floodplain Management Requirements: A Study Guide and Desk Reference for Local Officials*, National Flood Insurance Program, FEMA 480, Federal Emergency Management Agency, Washington, D.C.

FEMA, 2005c, *Promoting Seismic Safety: Guidance for Advocates*, FEMA 474, Federal Emergency Management Agency, Washington, D.C.

FEMA, 2005d, *Earthquake Safety Activities: For Children and Teachers*, FEMA 527, Federal Emergency Management Agency, Washington, D.C.

FEMA, 2006a, *Earthquake Preparedness: What Every Child Care Provider Needs to Know*, FEMA 240, Federal Emergency Management Agency, Washington, D.C.

FEMA, 2006b, *Hurricane Katrina in the Gulf Coast: Building Performance Observations, Recommendations, and Technical Guidance*, Mitigation Assessment Team Report, FEMA 549, Federal Emergency Management Agency, Washington, D.C.

FEMA, 2007, *Risk Management Series: Design Guide for Improving Critical Facility Safety from Flooding and High Winds*, FEMA P-543, Federal Emergency Management Agency, Washington, D.C.

FEMA, 2008, *Home Builder's Guide to Construction in Wildfire Zones*, FEMA P-737, Federal Emergency Management Agency, Washington, D.C.

FEMA, 2009a, *Hurricane Ike in Texas and Louisiana: Building Performance Observations, Recommendations, and Technical Guidance*, Mitigation Assessment Team Report, FEMA P-757, Federal Emergency Management Agency, Washington, D.C.

FEMA, 2009b, *Tornado Protection: Selecting Refuge Area in Buildings*, Second Edition FEMA P-431, Federal Emergency Management Agency, Washington, D.C.

FEMA, 2009c, *Unreinforced Masonry Buildings and Earthquakes*, FEMA P-774, Federal Emergency Management Agency, Washington, D.C.

FEMA, 2009d, *Vertical Evacuation from Tsunamis: A Guide for Community Officials*, FEMA P-646A, Federal Emergency Management Agency, Washington, D.C.

FEMA, 2010a, *Risk Management Series: Design Guide for Improving School Safety in Earthquakes, Floods, and High Winds*, FEMA P-424, Federal Emergency Management Agency, Washington, D.C.

FEMA, 2010b, *Substantial Improvement/Substantial Damage Desk Reference*, FEMA P-758, Federal Emergency Management Agency, Washington, D.C.

FEMA, 2011, *The Response to the 2011 Joplin, Missouri, Tornado, Lessons Learned Study*, Federal Emergency Management Agency, Washington, D.C.

FEMA, 2012a, *FEMA Teen CERT Guide*, Federal Emergency Management Agency, Washington, D.C. http://www.fema.gov/media-library -data/1449865324894-7898237eb0427d36e98932589825151b /teen_cert_launch_maintaintraining_508_111315.pdf, last accessed on August 16, 2016.

FEMA, 2012b, *Guidelines for Design of Structures for Vertical Evacuation from Tsunamis*, Second Edition, FEMA P-646, Federal Emergency Management Agency, Washington, D.C.

FEMA, 2012c, *Reducing the Risks of Nonstructural Earthquake Damage—A Practical Guide*, FEMA E-74, Federal Emergency Management Agency, Washington, D.C.

FEMA, 2012d, *Spring 2011 Tornadoes: April 25-28 and May 22: Building Performance Observations, Recommendations, and Technical Guidance*, Mitigation Assessment Team Report, FEMA P-908, Federal Emergency Management Agency, Washington, D.C.

FEMA, 2013a, *Floodproofing Non-Residential Buildings*, FEMA P-936, Federal Emergency Management Agency, Washington, D.C.

FEMA, 2013b, *Sample School Emergency Operations Plan*, Federal Emergency Management Agency, Washington, D.C.

FEMA, 2013c, *Snow Load Safety Guide*, FEMA P-957, Federal Emergency Management Agency, Washington, D.C.

FEMA, 2014, *Emergency Power Systems for Critical Facilities: A Best Practices Approach to Improving Reliability*, FEMA P-1019, Federal Emergency Management Agency, Washington, D.C.

FEMA, 2015a, *Case Study – School Community Safe Room: Wichita, Kansas*, Federal Emergency Management Agency, Washington, D.C.

FEMA, 2015b, *Rapid Visual Screening of Buildings for Potential Seismic Hazards: A Handbook*, Third Edition, FEMA P-154, Federal Emergency Management Agency, Washington, D.C.

FEMA, 2015c, *Safe Rooms for Tornadoes and Hurricanes: Guidance for Community and Residential Safe Rooms*, FEMA P-361, Federal Emergency Management Agency, Washington, D.C.

FEMA, 2016, *San Francisco Bay Area Coastal Study: Increased Flooding Scenario Maps*, Federal Emergency Management Agency, Washington, D.C.

FEMA, 2017a, "East Baton Rouge Parish School to receive $1.1 M more in assistance," *News Release*, NR-156, https://www.fema.gov/news-release/2017/05/11/east-baton-rouge-parish-schools-receive-11-m-more-assistance, last accessed on May 22, 2017.

FEMA, 2017b, *Hazus®: Estimated Annualized Earthquake Losses for the United States*, FEMA P-366, Federal Emergency Management Agency, Washington, D.C.

Fothergill, A., and Peek, L., 2015, *Children of Katrina,* University of Texas Press, Austin, Texas.

Freeman Health System, 2015, "Bill & Virginia Leffen Center for Autism celebrates completion of new campus," *Freeman News*, Joplin, Missouri, https://www.freemanhealth.com/about-us/freeman-news/bill-virginia-leffen-center-for-autism-celebrates-completion-of-new-campus, last accessed on February 22, 2017.

GADRRRES, 2017, *Comprehensive School Safety*, Global Alliance for Disaster Risk Reduction and Resilience in the Education Sector, United Nations International Strategy for Disaster Reduction.

Georgia Department of Education, 2012, *Crisis Management and Prevention Information for Georgia Public Schools*, Atlanta, Georgia.

GFDRR, 2009, *Guidance Notes on Safer School Construction*, developed as a collaboration between the InterAgency Network for Education in Emergencies (INEE) and the Global Facility for Disaster Reduction and Recovery (GFDRR), The World Bank.

Homeland Security, 2013, *Homeland Security Exercise and Evaluation Program (HSEEP)*, Washington, D.C.

ICC, 2000, *2000 International Building Code*, International Code Council, Inc., Country Club Hills, Illinois.

ICC, 2014a, *ICC/NSSA Standards for the Design and Construction of Storm Shelters*, ICC 500-2014, International Code Council & National Storm Shelter Association, Country Club Hills, Illinois.

ICC, 2014b, *2015 International Building Code*, International Code Council, Inc., Country Club Hills, Illinois.

IFCR, 2011, *Public Awareness and Public Education for Disaster Risk Reduction: A Guide*, International Federation of Red Cross and Red Crescent Societies, Geneva, Switzerland.

International Conference of Building Officials, 1976, *Uniform Building Code*, 1976 Edition, Whittier, California.

International Finance Corporation, 2010, *Disaster and Emergency Preparedness: Guidance for Schools*, World Bank Group, Washington, D.C.

Jaiswal, K.S., Petersen, M.D., Rukstales, K., and Leith, W.S., 2015, "Earthquake shaking hazard estimates and exposure changes in the conterminous United States," *Earthquake Spectra*, Vol. 31, No. S1, pp. S201-S220.

Kaiser, S., 2016, *Colorado School Safety: An Examination of Web Accessibility of Emergency Management Information*, M.A. Thesis, Department of Sociology, Colorado State University, Fort Collins, Colorado.

Kanter, R.K., and Abramson, D., 2013, "School interventions after the Joplin tornado," *Prehospital and Disaster Medicine*, Vol. 29, No. 2, pp. 214-217.

Lussier, C., 2017, "Baton Rouge school flood damage tops $60 million, but not all schools likely to be repaired," *The Advocate*, Baton Rouge, Louisiana, http://www.theadvocate.com/baton_rouge/news /education/article_81eb360c-e825-11e6-ad1a-bfc3dd3bc235.html, last accessed on March 2, 2017.

Magistrale, R., 2015, "Ocosta Elementary Week End Wrap Up 10-29-15," YouTube video, 2:07, filmed October 2015, https://youtu.be /G0AzsXTZhzw, last accessed on June 7, 2017.

Manning, R., 2012, "Portland parents worried about quake safety in schools," *Oregon Public Broadcasting*, March 8, 2012, http://www.opb.org /news/article/portland-parents-worried-about-quake-safety-schools/, last accessed on June 11, 2017.

McTavish, E., 2016, *Since Joplin Tornado, Dozens of Safe Rooms Built in Area Schools, Communities*, KSMU Radio, http://ksmu.org/post /joplin-tornado-dozens-safe-rooms-built-area-schools-communities #stream/0, last accessed on May 17, 2017.

Miller, D., 1960, *Giant Waves in Lituya Bay Alaska*, Geological Survey Professional Paper 354-C, United States Department of the Interior, United States Government Printing Office, Washington, D.C., https://pubs.usgs.gov/pp/0354c/report.pdf, last accessed on March 22, 2017.

Missouri Professional School Counselors and Counselor Educators, 2015, *School-Wide Crisis Management Plan Guide: A Professional School Counselor's Guide to School-Wide Crisis Management*, Comprehensive Guidance and Counseling Program: Responsive Services, Missouri.

Mitchell, C., 2016, "Scores of Louisiana school remain closed after severe flooding," *Education Week*, Bethesda, Maryland, http://blogs.edweek .org/edweek/District_Dossier/2016/08/scores_of_louisiana_schools _remain_closed.html, last accessed on May 23, 2017.

Multihazard Mitigation Council, 2005, *Natural Hazard Mitigation Saves: An Independent Study to Assess the Future Savings from Mitigation Activities*, Vol 1 – Findings, Conclusions, and Recommendations, prepared with funding from the Federal Emergency Management Agency of the U.S. Department of Homeland Security, National Institute of Building Sciences, Washington, D.C.

National Clearinghouse for Educational Facilities, 2008a, *Earthquakes and Schools*, National Institute of Building Sciences, prepared under a grant from the U.S. Department of Education, Washington, D.C.

National Clearinghouse for Educational Facilities, 2008b, *Wildfires and Schools*, National Institute of Building Sciences, prepared under a grant from the U.S. Department of Education, Washington, D.C.

National Education Association, 2015, *School Crisis Guide: Help and Healing in a Time of Crisis*, Health Information Network, Washington, D.C.

National Forum on Education Statistics, 2010, *Crisis Data Management: A Forum Guide to Collecting and Managing Data About Displaced Students*, NFES 2010-804, National Center for Education Statistics, U.S. Department of Education, Washington, D.C.

National Hurricane Center, 2016, *Storm Surge Overview*, National Oceanic and Atmospheric Administration, http://www.nhc.noaa.gov/surge/, last accessed on March 1, 2017.

NCES, 2016, "Fast facts: Back to school statistics," National Center for Education Statistics, Institute of Education Sciences, U.S. Department of Education, https://nces.ed.gov/fastfacts /display.asp?id=372, last accessed on February 20, 2017.

NISEE, 1935, *Karl V. Steinbrugge Collection*, NISEE/PEER/UCB, National Information Service for Earthquake Engineering/Pacific Earthquake Engineering Research Center/University of California, Berkeley, California, https://nisee.berkeley.edu/elibrary/Image/S2979, last accessed on February 28, 2017.

NISEE, 1949, *Karl V. Steinbrugge Collection*, NISEE/PEER/UCB, National Information Service for Earthquake Engineering/Pacific Earthquake Engineering Research Center/University of California, Berkeley, California, https://nisee.berkeley.edu/elibrary/Image/S2994, last accessed on February 28, 2017.

NIST, 2013, *Preliminary Reconnaissance of the May 20, 2013, Newcastle-Moore Tornado in Oklahoma*, NIST Special Publication 1164, National Institute of Standards and Technology, Gaithersburg, Maryland.

NIST, 2014, *Cost Analyses and Benefit Studies for Earthquake-Resistant Construction in Memphis, Tennessee*, NIST GCR 14-917-26, prepared by the NEHRP Consultants Joint Venture, a partnership of the Applied Technology Council and the Consortium for Universities for Research in Earthquake Engineering, for the National Institute of Standards and Technology, Gaithersburg, Maryland.

NOAA, 2014, "NOAA: 'Nuisance flooding' an increasing problem as coastal sea levels rise," National Oceanic and Atmospheric Administration, http://www.noaanews.noaa.gov/stories2014/20140728 _nuisanceflooding.html, last accessed March 2, 2017.

NOAA, 2015, "How many tropical cyclones have there been each year in the Atlantic basin? What years were the greatest and fewest seen?," Hurricane Research Division, Atlantic Oceanographic & Meteorological Laboratory, National Oceanic & Atmospheric Administration, http://www.aoml.noaa.gov/hrd/tcfaq/E11.html, last accessed on March 2, 2017.

NOAA, 2016a, "IBTraCs," International Best Track Archive for Climate Stewardship (IBTrACS), National Oceanic and Atmospheric Administration, National Centers for Environmental Information, Storm Prediction Center, https://www.ncdc.noaa.gov/ibtracs/, last accessed on March 20, 2017.

NOAA, 2016b, "SVRGIS," National Oceanic and Atmospheric Administration, National Weather Service, Storm Prediction Center, www.spc.noaa.gov/gis/svrgis/, last accessed on March 20, 2017.

NOAA, 2017, "Historical records and trends," National Oceanic and Atmospheric Administration website, https://www.ncdc.noaa.gov /climate-information/extreme-events/us-tornado-climatology/trends, last accessed on February 14, 2017.

NSPRA, 2011, "In the aftermath of a devastating tornado, Joplin Public Schools' staff showed 'a lot of heart' and a lot of PR savvy," National School Public Relations Association, Rockville, Maryland, http://www.nspra.org/e_network/november_2011_leadstory, last accessed on February 20, 2017.

NSPRA, 2016, *The Complete Crisis Communication Management Manual for Schools*, National School Public Relations Association, Rockville, Maryland.

NTHMP, 2015, *United States and Territories National Tsunami Hazard Assessment: Historical Record and Sources for Waves—Update*, National Tsunami Hazard Mitigation Program, nws.weather.gov /nthmp/documents/Tsunami_Assessment_2016Update.pdf, last accessed on June 11, 2017.

NTHMP, 2016, *U.S. Tsunami Hazard*, National Tsunami Hazard Mitigation Program, http://nws.weather.gov/nthmp/documents/ushazard.pdf, last accessed on March 3, 2017.

O'Connor, P., 2013, "Pedagogy of love and care: Shaken schools respond," *Disaster Prevention and Management*, Vol. 22, No. 5, pp. 425-433.

Patrick, L., Solecki, W., Jacob, K.H., Kunreuther, H., and Nordenson, G., 2015, "New York City Panel on Climate Change 2015 Report, Chapter 3: Static Coastal Flood Mapping," Building the Knowledge Base for Climate Resiliency, *Annals of the New York Academy of Sciences*, Vol. 1336, pp. 45-55.

Petal, 2008, *Disaster Prevention for Schools: Guidance for Education Sector Decision-Makers*, International Strategy for Disaster Reduction Thematic Platform for Knowledge and Education, United Nations Office for Disaster Risk Reduction (UNISDR), Geneva, Switzerland.

Petersen, M.D., Moschetti, M.P., Powers, P.M., Mueller, C.S., Haller, K.M., Frankel, A.D., Zeng, Y., Rezaeian, S., Harmsen, S.C., Boyd, O.S., Field, N., Chen, R., Rukstales, K.S., Luco, N., Wheeler, R.L., Williams, R.A., and Olsen, A.H., 2014, *Documentation for the 2014 Update of the United States National Seismic Hazard Maps*, Open-File Report 2014–1091, U.S. Geological Survey.

Petersen, M.D., Mueller, C.S., Moschetti, M.P., Hoover, S.M., Llenos, A.L., Ellsworth, W.L., Michael, A.J., Rubinstein, J.L., McGarr, A.F., and Rukstales, K.S., 2016, *2016 One-Year Seismic Hazard Forecast for the Central and Eastern United States from Induced and Natural Earthquakes*, Open-File Report 2016–1035, Version.1.1, U.S. Geological Survey.

Pew, 2017, *Flooding Threatens Public Schools Across the Country*, The Pew Charitable Trusts, Philadelphia, Pennsylvania.

Powitzy, C., 2009, "The storm as a teacher: Lessons in preparedness from Hurricanes Ike and Rita," *Educational Facilities Planner*, Vol. 44, Issue 1.

PrepareAthon, 2014, *Prepare Your People for Wildfire Safety: K-12 Schools*, America's PrepareAthon, Ready Campaign, Washington, D.C.

Ramirez, M., Kubicek, K., Peek-Asa, C., and Wong, M., 2009, "Accountability and assessment of emergency drill performance at schools," *Family & Community Health*, Vol. 32, No. 2, pp. 105-114.

Redlener, I., Grant, R., Abramson, D., and Johnson, D., 2008, *The 2008 American Preparedness Project: Why Parents May Not Heed Evacuation Orders and What Emergency Planners, Families and Schools Need to Know*, National Center for Disaster Preparedness, Columbia University and The Children's Health Fund, New York, New York, http://www.childrenshealthfund.org/sites/default /files/NCDP-CHF-Evacuation-White-Paper-Sep2008.pdf, last accessed on May 17, 2017.

Risk RED, 2009, *School Disaster Readiness: Lessons from the First Great Southern California ShakeOut*, 2nd Edition, Risk RED for Earthquake Country Alliance Los Angeles, Los Angeles, California, http://www .preventionweb.net/files/14873_RR2008SchoolReadinessReport.pdf, last accessed on May 17, 2017.

Romano, L., 2010, "Hurricane Camille (August 1969)," *Encyclopedia Virginia*, Virginia Foundation for the Humanities, www.encyclopediavirginia.org/Hurricane_Camille_August _1969#start_entry, last accessed on May 17, 2017.

Rubinstein, J.L., and Mahani, A.B., 2015, "Myths and facts on wastewater injection, hydraulic fracturing, enhanced oil recovery, and induced seismicity," *Seismological Research Letters*, Vol. 86, No. 4.

Saiyed, Z., Fatima, S., Izumoto, K., Robertson, I., Welliver, B., and Tremayne, H., 2017, "Advocacy for tsunami mitigation and preparedness of U.S. schools along the Pacific Rim," *Proceedings*, Paper No. 4924, 16th World Conference on Earthquake Engineering, Santiago, Chile.

SAMHSA, 2012, *Tips for Talking with and Helping Children and Youth Cope After a Disaster or Traumatic Event: A Guide for Parents, Caregivers, and Teachers*, Substance Abuse and Mental Health Services Administration, SMA12-4732, Rockville, Maryland.

Save the Children, 2012, *Is America Prepared to Protect Our Most Vulnerable Children in Emergencies?*, A National Report Card on Protecting Children during Disasters, Fairfield, Connecticut.

Save the Children, 2015, *Still at Risk: U.S. Children 10 Years After Hurricane Katrina*, 2015 National Report Card on Protecting Children in Disasters, Fairfield, Connecticut.

School Facilities and Organization, 2014, *Washington State K-12 Facilities Hazard Mitigation Plan*, Office of Superintendent of Public Instruction (OSPI), prepared by Goettel, K., and Dengel, R., Olympia, Washington.

Scranton Products, 2016, *Duralife Lockers at Danville Middle School: Helping to Improve the Educational Experience*, Scranton, Pennsylvania, http://www.scrantonproducts.com/projects /duralife-lockers-case-studies/scranton-products-duralife-lockers -prove-to-be-a-time-and-labor-saver-for-re-built-danville-middle -school/, last accessed on May 17, 2017.

SEFT Consulting Group, 2015, *Beaverton School District Resilience Planning for High School at South Cooper Mountain and Middle School at Timberland*, Beaverton, Oregon.

Selby, D., and Kagawa, F., 2012, *Disaster Risk Reduction in School Curricula: Case Studies from Thirty Countries*, United Nations Educational, Scientific and Cultural Organization (UNESCO)/United Nations Children's Fund (UNICEF), Geneva, Switzerland.

Selby, D., and Kagawa, F., 2014, *Towards a Learning Culture of Safety and Resilience: Technical Guidance for Integrating Disaster Risk Reduction in the School Curriculum*, United Nations Educational, Scientific and Cultural Organization (UNESCO)/United Nations Children's Fund (UNICEF).

Shaw, R., Takeuchi, Y., and Fernandez, G., 2012, *School Recovery: Lessons from Asia*, Graduate School of Global Environmental Studies, Kyoto University, International Environment and Disaster Management, Kyoto, Japan.

Shen, Y.-J., and Sink, C.A., 2002, "Helping elementary-age children cope with disasters," *Professional School Counseling*, Vol. 5, No. 5, pp. 322-330.

Smart Vent, 2016, *Wet Floodproofing Non-Residential Buildings*, Foundation Flood Vents, Pitman, New Jersey, http://smartvent.com/images/uploads/product_documents/smartvent-wetfloodproofing-nonres.pdf, last accessed on March 1, 2017.

Smith, A., 2015, *U.S. Smartphone Use in 2015*, Pew Research Center, http://www.pewinternet.org/2015/04/01/us-smartphone-use-in-2015/, last accessed on February 20, 2017.

Southern Education Foundation, 2015, "A new majority research bulletin: Low income students now a majority in the nation's public schools," *New Majority Report Series*, Atlanta, Georgia.

Stalker, S.L., Cullen, T., and Kloesel, K., 2015, "Using PBL to prepare educators for severe weather," *The Interdisciplinary Journal of Problem-Based Learning*, Vol. 9, Issue 2.

Stewart, S.R., 2011, *Tropical Cyclone Report: Tropical Storm Allison*, National Hurricane Center, http://www.nhc.noaa.gov/data/tcr/AL012001_Allison.pdf, last accessed on March 1, 2017.

Tennessee Office of Homeland Security, 2014, *Emergency Operations Planning: A Model for Schools and Businesses*, Nashville, Tennessee.

The Telegraph, 2005, "Girl, 10, used geography lesson to save lives," *The Telegraph*, http://www.telegraph.co.uk/news/1480192/Girl-10-used-geography-lesson-to-save-lives.html, last accessed on February 20, 2017.

Thompson, S., 2016, "Construction flaws and deviations found in tornado-damaged school, officials say," *The Dallas Morning News*, Dallas, Texas.

Tobin-Gurley, J., 2016, *Educational Continuity Following the 2013 Colorado Floods*, Department of Sociology, Colorado State University, Fort Collins, Colorado.

UCLA Center for Public Health and Disasters, 2004, *Head Start: Disaster Preparedness Workbook*, University of California, Los Angeles.

U.S. Department of Education, 2006, "Creating emergency management plans," *ERCM Express*, Vol. 2, Issue 8, Emergency Response and Crisis Management Technical Assistance Center, Office of Safe and Drug-Free Schools, Washington, D.C.

U.S. Department of Education, 2007, *Practical Information on Crisis Planning a Guide for Schools and Communities*, Office of Safe and Drug-Free Schools, Washington, D.C.

U.S. Department of Education, 2008, *A Guide to School Vulnerability Assessments: Key Principles for Safe Schools*, Office of Safe and Drug-Free Schools, Washington, D.C.

U.S. Department of Education, 2012, "Digest of educational statistics," National Center for Education Statistics, Institute of Education Sciences, http://nces.ed.gov/programs/digest/d14/tables/dt14_217.10.asp, last accessed on June 11, 2017.

U.S. Department of Education, 2013, *Guide for Developing High-Quality School Emergency Operations Plans*, Office of Elementary and Secondary Education, Office of Safe and Healthy Students, Washington, D.C.

U.S. Department of Energy, 2011, *Rebuilding it Better: Greensburg, Kansas*, prepared by the National Renewable Energy Laboratory, http://www.nrel.gov/docs/fy11osti/49315.pdf, last accessed on February 20, 2017.

U.S. General Accountability Office, 2007, *Emergency Management: Status of School Districts' Planning and Preparedness*, Washington, D.C.

USGS, 2012, *Dam-Breach Analysis and Flood-Inundation Mapping for Lakes Ellsworth and Lawtonka near Lawton, Oklahoma*, United States Geological Survey, https://pubs.usgs.gov/sir/2012/5026/SIR12-5026.pdf, last accessed on May 17, 2017.

Utah State Office of Education, 2013, *Emergency Preparedness Planning Guide for Utah Public Schools*, Salt Lake City, Utah.

Virginia Department of Emergency Management, 2008, *Hurricane Storm Surge Maps*, Southside East Regional Map, North Chesterfield, Virginia, http://www.vaemergency.gov/prepare-recover/threats/hurricane-storm-surge-maps/, last accessed on March 1, 2017.

Wachtendorf, T., Brown, B., and Nickle, M.C., 2008, "Big Bird, Disaster Masters, and high school students taking charge: The social capacities of children in disaster education," *Children, Youth and Environments*, Vol. 18, No. 1, pp. 456-469.

Walker, M., Whittle, R., Medd, W., Burningham, K., Moran-Ellis, J., and Tapsell, S., 2010, *Children and Young People 'After the Rain has Gone' – Learning Lessons for Flood Recovery and Resilience*, final project report for 'Children, Flood and Urban Resilience: Understanding children and young people's experience and agency in the flood recovery process,' Lancaster University, Lancaster, United Kingdom.

Washington Military Department, 2012, *Project Safe Haven: Tsunami Vertical Evacuation on the Washington Coast, Pacific County*, Olympia, Washington, http://mil.wa.gov/uploads/pdf/emergency -management/haz_safehavenreport_pacific.pdf, last accessed on March 22, 2017.

Weather Prediction Center, 2016a, *Hurricane Agnes – June 14-25, 1972*, National Centers for Environmental Prediction, National Oceanic and Atmospheric Administration, College Park, Maryland, http:// www.wpc.ncep.noaa.gov/tropical/rain/agnes1972.html, last accessed on March 2, 2017.

Weather Prediction Center, 2016b, *Hurricane Frances – September 3-11, 2004*, National Centers for Environmental Prediction, National Oceanic and Atmospheric Administration, College Park, Maryland, http://www.wpc.ncep.noaa.gov/tropical/rain/frances2004.html, last accessed on March 2, 2017.

Project Participants

FEMA Oversight

Mike Mahoney (Project Officer)
Federal Emergency Management Agency
400 C Street, SW, Suite 313
Washington, D.C. 20472

Andrew Herseth (Task Monitor)
Federal Emergency Management Agency
400 C Street, SW, Suite 313
Washington, D.C. 20472

ATC Management and Oversight

Jon A. Heintz (Program Executive, Program Manager)
Applied Technology Council
201 Redwood Shores Parkway, Suite 240
Redwood City, California 94065

Ayse Hortacsu (Project Manager)
Applied Technology Council
201 Redwood Shores Parkway, Suite 240
Redwood City, California 94065

Veronica Cedillos (Project Manager)
Applied Technology Council
201 Redwood Shores Parkway, Suite 240
Redwood City, California 94065

Project Technical Committee

Barry H. Welliver (Project Technical Director)
BHW Engineers
13623 S Bridle Trail Road
Draper, Utah 84020

Lori Peek
University of Colorado-Boulder
Department of Sociology, 327 Ketchum
Boulder, Colorado 80309

Suzanne Frew
The Frew Group
828 Carmel Avenue
Albany, California 94706

John Schelling
Lacey, Washington 98513

William T. Holmes
Structural Engineer
2600 La Cuesta
Oakland, California 94611

Thomas L. Smith
TLSmith Consulting Inc.
16681 Boswell Road
Rockton, Illinois 61072

Christopher P. Jones
Durham, North Carolina

Edward Wolf
3414 NE 18th Avenue
Portland, Oregon 97212

Project Review Panel

Ines Pearce (Chair)
Pearce Global Partners, Inc.
5419 Hollywood Blvd, Ste C-466
Los Angeles, California 90027

Jill Barnes
Los Angeles Unified School District
333 South Beaudry Avenue, 24th Floor
Los Angeles, California 90017

Victor Hellman
118 Admiral Court
Hampton, Virginia 23669

Andrew Kennedy*
University of Notre Dame
College of Engineering, 168 Fitzpatrick Hall
Notre Dame, Indiana 46556

Rebekah-Paci Green
516 High Street, MS 9095
Western Washington University
Bellingham, Washington 98225

Cindy Swearingen
P.O. Box 63
Beggs, Oklahoma 74421

*ATC Board Contact

Report Development Consultant

Laura Dwelley-Samant
San Francisco, California

Working Group

Lucy Carter
Colorado State University
Department of Sociology
Fort Collins, Colorado 80523

Shawna Bendeck
Colorado State University
Department of Sociology
Fort Collins, Colorado 80523

Scott Kaiser
Colorado State University
Department of Sociology
Fort Collins, Colorado 80523

Jacob Moore
Colorado State University
Department of Sociology
Fort Collins, Colorado 80523

Meghan Mordy
Colorado State University
Department of Sociology
Fort Collins, Colorado 80523

Katherine Murphy
Colorado State University
Department of Sociology
Fort Collins, Colorado 80523

Jennifer Tobin
Colorado State University
Department of Sociology
Fort Collins, Colorado 80523